I0470709

OM SHRI MAHAGANAPATHAYE NAMA:

OM POORNAMADA: POORNAMIDAM POORNAT
POORNAMUDACHYATE POORNASYA POORNAMADAYA
POORNAMEVAVASHISHYATE

OM SHOONYAMADA: SHOONYAMIDAM SHOONYAT
SHOONYAMUDACHYATE
SHOONYASYA SHOONYAMADAYA
SHOONYAMEVAVASHISHYATE

Seamless Infinity ~
Some Generalized Papers in
Elementary Number Theory

Narayanan
(Narayanan Raghunathan)

©2012 Narayanan Raghunathan

ISBN: 978–1489514356

Front cover art by Narayanan Raghunathan
Back cover photo courtesy Sateesh (Charles) Thittamangalam

Book design by Manu Krishnan
Cover design by Shyam Santhanam

Contents

Prologue ∼

Seamless Infinity ∼ Some Generalized Papers in Elementary Number Theory is a collection of sixteen papers which can be comprehended and specially appreciated by anybody who has done a first course in Number Theory ie. who knows his "funda-s" as they used to say at IIT Bombay. But these papers can also be appreciated by enthusiastic amateurs and even bright school students. Generally most Mathematics papers of recent times are hardly read by anybody. Needless to say it applies to other subjects too.

Actually all these papers give out the results without explicit proof via.the universal process of mathematical induction. If I were to expand these papers fully with elaborate proofs, the volume of the book would be ten times, that is nearly two thousand pages! Mathematicians can surely fill up the details of the inductive proofs by themselves.

I observe I intend to publish a few monographs in Elementary Number Theory where I earnestly hope all the missing details in these papers would be added.

Here is a tentative list of the books to be published.

1) Generalized Results in Elementary Number Theory.
2) Multinomial Theorems and Essential Generalizations.
3) Triangular Number ∼ An Exhaustive Monograph.
4) Congruence Relations ∼ A Monograph.
5) Some Arithmetic Functions and Generalizations.
6) Infinite Families of Base Systems.
7) Infinite Generalized Fibonacci Sequences and Polynomials.
8) Polygonal Numbers ∼ A Monograph.
9) Infinite Cosmoses of Infinite Rhythmic Continued Fractions.

I don't personally think that these papers contained in this book need any special introduction. They stand out self evidently by their intrinsic inevitability.

In an age of statistical junk, fuzzy fraudulence and transfinite lies, these beautiful simple results need to be noted at least for their global aesthetic appeal and their primal arithmetic validity.

I profusely thank Manu Krishnan and Shyam Santhanam, for their kind assistance in preparing this book. I also thank my sister Prema Anand for helping me with the the proof reading.

Generalized Arithmetic Progressions – A Survey

Dedicated to my esteemed teacher Shri. S. Iyer who taught me Euclid, Descartes, Progressions.

Abstract. Given any Arithmetic Progression we could treat the Sequence of its Sum upto n terms as another Progression and Find a Formula for the Sum of this Progression, repeat the process as many times as we may desire {find the Super-Sums of the Original Arithmetic Progression upto any [α-Level] we may say} and Find a Formula for the Sum upto n terms in each case. The Progressive Rhythm upto any level [α- Level] may be cumulatively collected and A Generalized Arithmetic Progression Defined and its Sum upto n terms and Super-Sums upto n terms may also be formulated analytically.

Key Words:- Arithmetic Progressions, Infinite Generalizations.
AMS Subject Classification No:- 11A25; 11A99.

1. Super-sums of an Arithmetic Progression

Let $\begin{bmatrix} \mathbf{A}_n \\ [a,b] \end{bmatrix} = a + (n-1)b = \overset{[1]}{\underset{[a,b]}{\mathbf{A}_n}}$ be a given Arithmetic Progression with "a" as the Initiating Term and "b" as the Basic Difference where "a" and "b" are algebraic numbers. We can determine the Super-Sums upto any Level $\infty = 1, 2, 3 \cdots \infty$. The ordinary sum of the Arithmetic Progression is clearly the Super-Sum Level–1.

Notation 1.1. In $[S]_n \begin{bmatrix} [\alpha] \\ \mathbf{A}_n \\ [a,b] \end{bmatrix}$, $[S]_n$ is the Sum upto n terms of the αth Level of the Arithmetic Progression [super-sum α^{th} Level] and it yields the $\overset{[\alpha+1]}{\underset{[a,b]}{\mathbf{A}_n}}$ the $(\alpha+1)^{\text{th}}$ Level of the given Arithmetic Progression.

We define Arithmetic Progression.

$$\begin{bmatrix} \mathbf{A}_n \\ [a,b] \end{bmatrix} = a + (n-1)b = \overset{[1]}{\underset{[a,b]}{\mathbf{A}_n}}$$

ALL the following Sequence of Formulae could be easily Proved by the Method of Mathematical Induction by now traditionally formalized. For each Level, the induction is performed on "n" and for the General Result the induction is performed on "α". The Routine Steps are omitted to save Eternal Space-Time!

$$[S]_n \begin{bmatrix} [1] \\ \mathbf{A}_n \\ [a,b] \end{bmatrix} = an + \left[\frac{(n-1)(n)}{2\,!} \right] b = \begin{matrix} [2] \\ \mathbf{A}_n \\ [a,b] \end{matrix}$$

[Traditional Ancient Result]

$$[S]_n \begin{bmatrix} [2] \\ \mathbf{A}_n \\ [a,b] \end{bmatrix} = a\left[\frac{n(n+1)}{2} \right] + \left[\frac{(n-1)(n)(n+1)}{3\,!} \right] b = \begin{matrix} [3] \\ \mathbf{A}_n \\ [a,b] \end{matrix}$$

$$[S]_n \begin{bmatrix} [3] \\ \mathbf{A}_n \\ [a,b] \end{bmatrix} = a\left[\frac{n(n+1)(n+2)}{3\,!} \right]$$

$$+ \left[\frac{(n-1)(n)(n+1)(n+2)}{4\,!} \right] b = \begin{matrix} [4] \\ \mathbf{A}_n \\ [a,b] \end{matrix}$$

$$[S]_n \begin{bmatrix} [4] \\ \mathbf{A}_n \\ [a,b] \end{bmatrix} = a\left[\frac{n(n+1)(n+2)(n+3)}{4\,!} \right]$$

$$+ \left[\frac{(n-1)(n)(n+1)(n+2)(n+3)}{5\,!} \right] b$$

$$= \begin{matrix} [5] \\ \mathbf{A}_n \\ [a,b] \end{matrix} \ldots\ldots\ldots >>>>>>>>$$

$$[S]_n \begin{bmatrix} [\alpha-1] \\ \mathbf{A}_n \\ [a,b] \end{bmatrix} = a\left[\frac{n(n+1)(n+2)\text{------}(n+\alpha-2)}{(\alpha-1)\,!} \right]$$

$$+ \left[\frac{(n-1)(n)(n+1)\text{------}(n+\alpha-2)}{\alpha\,!} \right] b = \begin{matrix} [\alpha] \\ \mathbf{A}_n \\ [a,b] \end{matrix}$$

$$[S]_n \begin{bmatrix} [\alpha] \\ \mathbf{A}_n \\ [a,b] \end{bmatrix} = a\left[\frac{n(n+1)(n+2)\text{------}(n+\alpha-1)}{\alpha\,!} \right]$$

$$+ \left[\frac{(n-1)(n)(n+1)\text{------}(n+\alpha-1)}{(\alpha+1)\,!} \right] b = \begin{matrix} [\alpha+1] \\ \mathbf{A}_n \\ [a,b] \end{matrix}$$

$$[S]_n \begin{bmatrix} [\alpha + 1] \\ \mathbf{A}_n \\ [a, b] \end{bmatrix} = a \left[\frac{n(n+1)(n+2) \text{------}(n+\alpha)}{(\alpha+1)!} \right]$$

$$+ \left[\frac{(n-1)(n)(n+1) \text{------}(n+\alpha)}{(\alpha+2)!} \right] b$$

$$= \begin{matrix} [\alpha + 2] \\ \mathbf{A}_n \\ [a, b] \end{matrix} \ \text{----} >>>> \infty\infty \text{ FOR EVER } \infty$$

2. Inductive-progression

$$\begin{bmatrix} \mathbf{I}_n \\ [a, b, c] \end{bmatrix} = a + (n-1)b + \left[\frac{(n-2)(n-1)}{2} \right] c = \begin{matrix} [1] \\ \mathbf{I}_n \\ [a, b, c] \end{matrix}$$

"a" as the Initiating Term and "b" as the Basic Difference and "c" is the inductive Factor [a,b, and c are algebraic numbers]. We can determine the Super-Sums upto any Level $\alpha = 1, 2, 3 \text{---}\infty$. The ordinary Sum of the INDUCTIVE-PROGRESSION IS clearly the Super-Sum Level - 1.

Notation 2.1. In $[S]_n \begin{bmatrix} [\alpha] \\ \mathbf{I}_n \\ [a, b, c] \end{bmatrix}$ $[S]_n$ is the Sum upto n terms of the αth Level of the INDUCTIVE-PROGRESSION [super-sum α^{th} Level] and it yields the $\begin{matrix} [a+1] \\ \mathbf{I}_n \\ [a, b, c] \end{matrix}$ the $(\alpha + 1)^{\text{th}}$ Level of the given INDUCTIVE- PROGRESSION.

$$\begin{bmatrix} \mathbf{I}_n \\ [a, b, c] \end{bmatrix} = a + (n-1)b + \left[\frac{(n-2)(n-1)}{2} \right] c = \begin{matrix} [1] \\ \mathbf{I}_n \\ [a, b, c] \end{matrix}$$

{An interesting example of INDUCTIVE-PROGRESSION is the set of Triangular Numbers. [Substitute $a = 1, b = 2, c = 1$].}

Now we can easily prove by Induction that the sum of this progression up to n terms is

$$[S]_n \begin{bmatrix} [1] \\ \mathbf{I}_n \\ [a, b, c] \end{bmatrix} = an + \left[\frac{(n-1)(n)}{2!} \right] b + \left[\frac{(n-2)(n-1)(n)}{3!} \right] c = \begin{matrix} [2] \\ \mathbf{I}_n \\ [a, b, c] \end{matrix}$$

ALL the following Sequence of Formulae could be easily Proved by the Method of Mathematical Induction by now traditionally formalized. For each Level, the induction is performed on "n" and for the General Result the induction is performed on "α". The Routine Steps are omitted to save Eternal Space-Time!

$$[S]_n \begin{bmatrix} \overset{[2]}{\mathbf{I}_n} \\ [a,b,c] \end{bmatrix} = a \left[\frac{n(n+1)}{2!} \right] + \left[\frac{(n-1)(n)(n+1)}{3!} \right] b$$

$$+ \left[\frac{(n-2)(n-1)(n)(n+1)}{4!} \right] c = \begin{matrix} \overset{[3]}{\mathbf{I}_n} \\ [a,b,c] \end{matrix}$$

$$[S]_n \begin{bmatrix} \overset{[3]}{\mathbf{I}_n} \\ [a,b,c] \end{bmatrix} = a \left[\frac{n(n+1)(n+2)}{3!} \right] + \left[\frac{(n-1)(n)(n+1)(n+2)}{4!} \right] b$$

$$+ \left[\frac{(n-2)(n-1)(n)(n+1)(n+2)}{5!} \right] c = \begin{matrix} \overset{[4]}{\mathbf{I}_n} \\ [a,b,c] \end{matrix}$$

$$[S]_n \begin{bmatrix} \overset{[4]}{\mathbf{I}_n} \\ [a,b,c] \end{bmatrix} = a \left[\frac{n(n+1)(n+2)(n+3)}{4!} \right]$$

$$+ \left[\frac{(n-1)(n)(n+1)(n+2)(n+3)}{5!} \right] b$$

$$+ \left[\frac{(n-2)(n-1)(n)(n+1)(n+2)(n+3)}{6!} \right] c = \begin{matrix} \overset{[5]}{\mathbf{I}_n} \\ [a,b,c] \end{matrix}$$

$$- - - - - - - - >>>>>> \infty\infty \text{ FOR EVER } \infty$$

$$[S]_n \begin{bmatrix} \overset{[\alpha-1]}{\mathbf{I}_n} \\ [a,b,c] \end{bmatrix} = a \left[\frac{n(n+1)(n+2) - - - - - -(n+\alpha-2)}{(\alpha-1)!} \right]$$

$$+ \left[\frac{(n-1)(n)(n+1) - - - - - -(n+\alpha-2)}{\alpha!} \right] b$$

$$+ \left[\frac{(n-2)(n-1)(n) - - -(n+\alpha-2)}{(\alpha+1)!} \right] c = \begin{matrix} \overset{[\alpha]}{\mathbf{I}_n} \\ [a,b,c] \end{matrix}$$

$$[S]_n \begin{bmatrix} \overset{[\alpha]}{\mathbf{I}_n} \\ [a,b,c] \end{bmatrix} = a \left[\frac{n(n+1)(n+2) - - - - - -(n+\alpha-1)}{\alpha!} \right]$$

$$+ \left[\frac{(n-1)(n)(n+1) - - - - - -(n+\alpha-1)}{(\alpha+1)!} \right] b$$

$$+ \left[\frac{(n-2)(n-1)(n) - - -(n+\alpha-1)}{(\alpha+2)!} \right] c = \begin{matrix} \overset{[\alpha+1]}{\mathbf{I}_n} \\ [a,b,c] \end{matrix}$$

$$[S]_n \begin{bmatrix} [\alpha + 1] \\ \mathbf{I}_n \\ [a, b, c] \end{bmatrix} = a \left[\frac{n(n+1)(n+2)\text{-----}(n+\alpha)}{(\alpha+1)\,!} \right]$$

$$+ \left[\frac{(n-1)(n)(n+1)\text{-----}(n+\alpha)}{(\alpha+2)\,!} \right] b$$

$$+ \left[\frac{(n-2)(n-1)(n)\text{---}(n+\alpha)}{(\alpha+3)\,!} \right] c = \begin{array}{c} [\alpha+2] \\ \mathbf{I}_n \\ [a, b] \end{array}$$

$$-------\ggg\ggg \infty\infty \text{ FOR EVER } \infty$$

3. Generalized Arithmetic Progressions-simple [* Arithmetic Progression]

We define the
GENERALIZED ARITHMETIC PROGRESSIONS-
SIMPLE [* ARITHMETIC PROGRESSION]

$$\begin{bmatrix} A^*_n \\ [a, b_1, \text{ ---}, b_\beta] \end{bmatrix} = a + (n-1)b_1 + \left[\frac{(n-1)(n)}{2} \right] b_2$$

$$+ \left[\frac{(n-1)(n)(n+1)}{3\,!} \right] b_3 + \text{---}+$$

$$+ \left[\frac{(n-1)(n)(n+1)\text{---}(n+\beta-2)}{\beta\,!} \right] b_\beta$$

$$= a + \sum_{\phi=1}^{\beta} \left[\frac{(n-1)\text{---}(n+\phi-2)}{\phi\,!} \right] b_\phi = \begin{array}{c} [1] \\ \mathbf{A}^*_n \\ [a, b_1, \text{ ---}, b_\beta] \end{array}$$

"a" as the initiating Term and "b_1", "b_2", "b_3" ---"b_β" as the Basic Differences [$a, b_1, b_2, b_3, \text{ ---}, b_\beta$ are algebraic numbers]. We can determine the Super-Sums of the GENERALIZED ARITHMETIC PROGRESSION -SIMPLE UPTO any Level $\alpha = 1, 2, 3\text{---}\infty$. The ordinary Sum of GENERALIZED ARITHMETIC PROGRESSIO -SIMPLE is clearly the Super-Sum Level-1 of the same. "β" may be called Generalization of the GENERALIZED ARITHMETIC PROGRESSION -SIMPLE.

Notation 3.1. In $[S]_n \begin{bmatrix} [\alpha] \\ \mathbf{A}^*_n \\ [a, b_1, \text{ ---}, b_\beta] \end{bmatrix}$ $[S]_n$ is the Sum upto n terms of the αth Level of the GENERALIZED ARITHMETIC PROGRESSION-SIMPLE [Super-Sum α^{th} Level] and it yields the $\begin{array}{c} [\alpha+1] \\ \mathbf{I}_n \\ [a, b, c] \end{array}$ the $(\alpha+1)^{\text{th}}$ Level of the given GENERALIZED ARITHMETIC PROGRESSION-SIMPLE.

$$\begin{bmatrix} \overset{*}{\mathbf{A}}_n \\ [a, b_1, \text{---}, b_\beta] \end{bmatrix} = a + \sum_{\phi=1}^{\beta} \left[\frac{(n-1)\text{---}(n+\phi-2)}{\phi!} \right] b_\phi$$

$$= \begin{bmatrix} [\alpha] \\ \overset{*}{\mathbf{A}}_n \\ [a, b_1, \text{---}, b_\beta] \end{bmatrix}$$

ALL the following Sequence of Formulae could be easily Proved by the Method of Mathematical Induction by now traditionally formalized. For each Level, the induction is performed on "n"and for The General Result the induction is performed on "α". The Routine Steps are omitted to save Eternal Space -Time!

$$[S]_n \begin{bmatrix} [1] \\ \overset{*}{\mathbf{A}}_n \\ [a, b_1, \text{---}, b_\beta] \end{bmatrix} = an + \left[\frac{(n-1)(n)}{2!} \right] b_1$$

$$+ \left[\frac{(n-1)(n)(n+1)}{3!} \right] b_2 + \text{---}+$$

$$+ \left[\frac{(n-1)(n)(n+1)\text{---}(n+\beta-1)}{(\beta+1)!} \right] b_\beta$$

$$= an + \sum_{\phi=1}^{\beta} \left[\frac{(n-1)\text{---}(n+\phi-1)}{(\phi+1)}(\phi+1)! \right] b_\phi$$

$$= \begin{bmatrix} [2] \\ \overset{*}{\mathbf{A}}_n \\ [a, b_1, \text{---}, b_\beta] \end{bmatrix}$$

$$[S]_n \begin{bmatrix} [2] \\ \overset{*}{\mathbf{A}}_n \\ [a, b_1, \text{---}, b_\beta] \end{bmatrix} = a \left[\frac{n(n+1)}{2!} \right] + \sum_{\phi=1}^{\beta} \left[\frac{(n-1)\text{---}(n+\phi)}{(\phi+2)!} \right] b_\phi$$

$$= \begin{array}{c} [3] \\ \overset{*}{\mathbf{A}}_n \\ [a, b_1, \text{---}, b_\beta] \end{array}$$

$$[S]_n \begin{bmatrix} [3] \\ \overset{*}{\mathbf{A}}_n \\ [a, b_1, \text{---}, b_\beta] \end{bmatrix} = a \left[\frac{n(n+1)(n+2)}{3!} \right]$$

$$+ \sum_{\phi=1}^{\beta} \left[\frac{(n-1)\text{---}(n+\phi+1)}{(\phi+3)!} \right] b_\phi$$

$$= \begin{array}{c} [4] \\ \overset{*}{\mathbf{A}}_n \\ [a, b_1, \text{---}, b_\beta] \end{array}$$

$$[S]n \begin{bmatrix} [4] \\ \mathbf{A}^*_n \\ [a, b_1, \text{---}, b_\beta] \end{bmatrix} = a \left[\frac{n(n+1)(n+2)(n+3)}{4!} \right]$$

$$+ \sum_{\phi=1}^{\beta} \left[\frac{(n-1)\text{---}(n+\phi+2)}{(\phi+4)!} \right] b_\phi$$

$$= \begin{matrix} [5] \\ \mathbf{A}^*_n \\ [a, b_1, \text{---}, b_\beta] \end{matrix}$$

$$- - - - - - - - >>>>>>>>>>>>>>>>>>>$$

$$[S]n \begin{bmatrix} [\alpha-1] \\ \mathbf{A}^*_n \\ [a, b_1, \text{---}, b_\beta] \end{bmatrix} = a \left[\frac{n(n+1)(n+2)\text{---}(n+\alpha-2)}{(\alpha-1)!} \right]$$

$$+ \sum_{\phi=1}^{\beta} \left[\frac{(n-1)\text{---}(n+\phi+\alpha-3)}{(\phi+\alpha-1)!} \right] b_\phi$$

$$= \begin{matrix} [\alpha] \\ \mathbf{A}^*_n \\ [a, b_1, \text{---}, b_\beta] \end{matrix}$$

$$[S]n \begin{bmatrix} [\alpha] \\ \mathbf{A}^*_n \\ [a, b_1, \text{---}, b_\beta] \end{bmatrix} = a \left[\frac{n(n+1)(n+2)\text{---}(n+\alpha-1)}{\alpha!} \right]$$

$$+ \sum_{\phi=1}^{\beta} \left[\frac{(n-1)\text{---}(n+\phi+\alpha-2)}{(\phi+\alpha)!} \right] b_\phi$$

$$= \begin{matrix} [\alpha+1] \\ \mathbf{A}^*_n \\ [a, b_1, \text{---}, b_\beta] \end{matrix}$$

$$[S]n \begin{bmatrix} [\alpha+1] \\ \mathbf{A}^*_n \\ [a, b_1, \text{---}, b_\beta] \end{bmatrix} = a \left[\frac{n(n+1)(n+2)\text{---}(n+\alpha)}{(\alpha+1)!} \right]$$

$$+ \sum_{\phi=1}^{\beta} \left[\frac{(n-1)\text{---}(n+\phi+\alpha-1)}{(\phi+\alpha+1)!} \right] b_\phi$$

$$= \begin{matrix} [\alpha+2] \\ \mathbf{A}^*_n \\ [a, b_1, \text{---}, b_\beta] \end{matrix}$$

$$- - - - - - - - >>>>>> \infty\infty \text{ FOR EVER } \infty$$

4. Generalized Arithmetic Progressions-Random [$*^r$ Arithmetic Progressions]

We define the
GENERALIZED ARITHMETIC PROGRESSIONS -RANDOM
[$*^r$ ARITHMETIC PROGRESSIONS]

$$\begin{bmatrix} \mathbf{A}_n^{*r} \\ [a, b_1, ---, b_\beta] \\ [@, j_1, ---, j_\beta] \end{bmatrix} = a + (n - j_1)b_1 + \left[\frac{(n - j_2)(n - j_2 + 1)}{2!} \right] b_2$$

$$+ \left[\frac{(n - j_3)(n - j_3 + 1)(n - j_3 + 2)}{3!} \right] b_3 + \cdots$$

$$+ \left[\frac{(n - j_\beta)(n - j_\beta + 1)(n - j_\beta + 2) \cdots (n - j_\beta + \beta - 1)}{\beta!} \right] b_\beta$$

$$= a + \sum_{\phi=1}^{\beta} \left[\frac{(n - j_\phi) \cdots (n - j_\phi + \phi - 1)}{\phi!} \right] b_\phi$$

$$= \begin{matrix} [1] \\ \mathbf{A}_n^{*r} \\ [a, b_1, ---, b_\beta] \\ [@, j_1, ---, j_\beta] \end{matrix}$$

with "a" as the Initiating Term and "b_1", "b_2","b_3" --- "b_β" as the Basic Differences [$a, b_1, b_2, b_3, ---, b_\beta$ are algebraic numbers] induced at [@] "j_1", "j_2", "j_2", ---,"j_β" respectively [$j_i > 1$, $i = 1, 2 --- \beta$]. We can determine the Super-Sums of the GENERALIZED ARITHMETIC PROGRESSION-RANDOM upto any Level $\alpha = 1, 2, --- \infty$. The ordinary Sum of GENERALIZED ARITHMETIC PROGRESSION -RANDOM is clearly the Super- Sum Level -1 of the same. "β" may be called Generalization- Level of the GENERALIZED ARITHMETIC PROGRESSION -RANDOM. J_i, ($I = 1, 2 --- \beta$) may be called the Inductive Points of the GENERALIZED ARITHMETIC PROGRESSION -RANDOM.

Notation 4.1. In $[S]_n \begin{bmatrix} [\alpha] \\ \mathbf{A}_n^{*r} \\ [a, b_1, ---, b_\beta] \\ [@ j_1, ---, j_\beta] \end{bmatrix}$, $[S]_n$ is the sum upto n terms of the

α^{th} Level of the GENERALIZED ARITHMETIC PROGRESSION-RANDOM

[Super-Sum α^{th} Level] and it yields the $\begin{matrix} [a+1] \\ \mathbf{A}_n^{*r} \\ [a, b_1, ---, b_\beta] \\ [@ j_1, ---, j_\beta] \end{matrix}$ the $(\alpha + 1)^{\text{th}}$ Level of the

given GENERALIZED ARITHMETIC PROGRESSION-RANDOM.

$$\begin{bmatrix} \mathbf{A}_n^{*r} \\ [a, b_1, ---, b_\beta] \\ [@, j_1, ---, j_\beta] \end{bmatrix} = a + \sum_{\phi=1}^{\beta} \left[\frac{(n - j_\phi) --- (n - j_\phi + \phi - 1)}{\phi!} \right] b_\phi$$

$$[1]$$

$$= \frac{\mathbf{A}_n^{*r}}{\begin{array}{c}[a, b_1, ---, b_\beta] \\ [@j_1, ---, j_\beta]\end{array}}.$$

ALL the following Sequence of Formulae could be easily Proved by the Method of Mathematical Induction by now traditionally formalized. For each Level, the induction is performed on "n" and for The General Result the induction is performed on "α". The Routine Steps are omitted to save Eternal Space-Time!

$$\begin{bmatrix} [1] \\ \mathbf{A}_n^{*r} \\ [a, b_1, ---, b_\beta] \\ [@, j_1, ---, j_\beta] \end{bmatrix} = an + \sum_{\phi=1}^{\beta} \left[\frac{(n - j_\phi) --- (n - j_\phi + \phi)}{(\phi + 1)!} \right] b_\phi$$

$$[2]$$

$$= \frac{\mathbf{A}_n^{*r}}{\begin{array}{c}[a, b_1, ---, b_\beta] \\ [@j_1, ---, j_\beta]\end{array}}$$

$$[S]_n \begin{bmatrix} [2] \\ \mathbf{A}_n^{*r} \\ [a, b_1, ---, b_\beta] \\ [@, j_1, ---, j_\beta] \end{bmatrix} = a \left[\frac{n(n+1)}{2!} \right]$$

$$+ \sum_{\phi=1}^{\beta} \left[\frac{(n - j_\phi) --- (n - j_\phi + \phi + 1)}{(\phi + 2)!} \right] b_\phi$$

$$[3]$$

$$= \frac{\mathbf{A}_n^{*r}}{\begin{array}{c}[a, b_1, ---, b_\beta] \\ [@j_1, ---, j_\beta]\end{array}}$$

$$[S]_n \begin{bmatrix} [3] \\ \mathbf{A}_n^{*r} \\ [a, b_1, ---, b_\beta] \\ [@, j_1, ---, j_\beta] \end{bmatrix} = a \left[\frac{n(n+1)(n+2)}{3!} \right]$$

$$+ \sum_{\phi=1}^{\beta} \left[\frac{(n - j_\phi) --- (n - j_\phi + \phi + 2)}{(\phi + 3)!} \right] b_\phi$$

$$[4]$$

$$= \frac{\mathbf{A}_n^{*r}}{\begin{array}{c}[a, b_1, ---, b_\beta] \\ [@j_1, ---, j_\beta]\end{array}}$$

$$[S]_n \begin{bmatrix} \overset{[4]}{\mathbf{A}}{}^{*r}_n \\ [a, b_1, ---, b_\beta] \\ [@, j_1, ---, j_\beta] \end{bmatrix} = a \left[\frac{n(n+1)(n+2)(n+3)}{4!} \right]$$

$$+ \sum_{\phi=1}^{\beta} \left[\frac{(n-j_\phi)---(n-j_\phi+\phi+3)}{(\phi+4)!} \right] b_\phi$$

$$= \overset{[5]}{\mathbf{A}}{}^{*r}_n \\ [a, b_1, ---, b_\beta] \\ [@j_1, ---, j_\beta]$$

$- - - - - - - >>>>>>>>>>>>>>>>>>>$

$$[S]_n \begin{bmatrix} \overset{[\alpha-1]}{\mathbf{A}}{}^{*r}_n \\ [a, b_1, ---, b_\beta] \\ [@, j_1, ---, j_\beta] \end{bmatrix} = a \left[\frac{n(n+1)(n+2)\cdots(n+\alpha-2)}{(\alpha-1)!} \right]$$

$$+ \sum_{\phi=1}^{\beta} \left[\frac{(n-j_\phi)---(n-j_\phi+\phi+\alpha-2)}{(\phi+\alpha-1)!} \right] b_\phi$$

$$= \overset{[\alpha]}{\mathbf{A}}{}^{*r}_n \\ [a, b_1, ---, b_\beta] \\ [@j_1, ---, j_\beta]$$

$$[S]_n \begin{bmatrix} \overset{[\alpha]}{\mathbf{A}}{}^{*r}_n \\ [a, b_1, ---, b_\beta] \\ [@, j_1, ---, j_\beta] \end{bmatrix} = a \left[\frac{n(n+1)(n+2)\cdots(n+\alpha-1)}{\alpha!} \right]$$

$$+ \sum_{\phi=1}^{\beta} \left[\frac{(n-j_\phi)\cdots(n-j_\phi+\phi+\alpha-1)}{(\phi+\alpha)!} \right] b_\phi$$

$$= \overset{[\alpha+1]}{\mathbf{A}}{}^{*r}_n \\ [a, b_1, ---, b_\beta] \\ [@j_1, ---, j_\beta]$$

$$[S]n \begin{bmatrix} [\alpha + 1] \\ \mathbf{A}_n^{*r} \\ [a, b_1, \cdots, b_\beta] \\ [@, j_1, \text{---}, j_\beta] \end{bmatrix} = a \left[\frac{n(n+1)(n+2) \cdots (n+\alpha)}{(\alpha + 1)} \right]$$

$$+ \sum_{\phi=1}^{\beta} \left[\frac{(n - j_\phi) \cdots (n - j_\phi + \phi + \alpha + 1)}{(\phi + \alpha + 1)!} \right] b_\phi$$

$$= \begin{matrix} [\alpha + 2] \\ \mathbf{A}_n^{*r} \\ [a, b_1, \text{---}, b_\beta] \\ [@j_1, \text{---}, j_\beta] \end{matrix}$$

$\cdots >>> \infty\infty$ FOR EVER ∞

We could of course induct and repeat the same inductive-block at as many random points of entry and generalize appropriately. The details though trivial are cumbersome. [1]

Since the Initiating term and Basic differences can be any algebraic number, we can see that each family of the Progressions elucidated here, defines a unique Algebraic Field of Sequences.

References

[1] Narayanan Raghunathan: *Functions and their Progressions- An Elementary Text.* [unpublished]

Generalized Geometric Progressions – A Survey

(Dedicated to my esteemed teacher Shri. C. G. George who taught me Algebra, Calculus (SSKZM))

Abstract. Given any Geometric Progression we could treat the Sequence of its Sum upto n terms as another Progression and Find a Formula for the Sum of this Progression, repeat the process as many times as we may desire {find the Super–Sums of the Original Geometric Progression upto any [α–Level] we may say} and Find a Formula for the Sum upto n terms in each case. The Progressive Rhythm upto any level [α–Level] may be cumilatively collected and A Generalized Geometric Progression Defined and its Sum upto n terms and Super–Sums upto n terms may also be formulated analytically.

Key Words:- Geometric Progressions, Infinite Generalizations
AMS Subject Classification No:- 11A99

1. Super–Sums of a Geometric Progression

Let $\begin{bmatrix} G_n \\ [a,r] \end{bmatrix} = ar^{(n-1)} = \overset{[1]}{\underset{[a,r]}{G_n}}$ $[n \geq 1]$ be a given Geometric Progression with "a" as the Initiating Term and "r" as the Common Ratio where "a" and "r" are algebraic numbers $(r \neq 1)$. We can determine the Super–Sums upto any Level $\alpha = 1, 2, 3$ --- ∞. The ordinary Sum of the Geometric Progression is clearly the Super–Sum Level–1.

Notation 1.1. In $[S]_n \begin{bmatrix} [\alpha] \\ G_n \\ [a,r] \end{bmatrix}$, $[S]_n$ is the Sum upto n terms of the αth Level of

the Geometric Progression [Super–Sum α^{th} Level] and it yields the $\begin{bmatrix} [\alpha+1] \\ G_n \\ [a,r] \end{bmatrix}$

the $(\alpha + 1)^{th}$ Level of the given Geometric Progression.

We define the Geometric Progression.

$$\begin{bmatrix} G_n \\ [a,r] \end{bmatrix} = ar^{(n-1)} = \overset{[1]}{\underset{[a,r]}{G_n}}$$

ALL the following Sequence of Formulae could be easily Proved by the Method of Mathematical Induction by now traditionally formalized. For each Level, the induction is performed on "n" and for The General Result the induction is performed on "α". The Routine Steps are omitted to save Eternal Space–Time !

$$[S]_n \begin{bmatrix} [1] \\ G_n \\ [a,r] \end{bmatrix} = \frac{a(r^n - 1)}{r - 1} = \begin{matrix} [2] \\ G_n \\ [a,r] \end{matrix} \qquad [\text{ Traditional Ancient Result }]$$

$$[S]_n \begin{bmatrix} [2] \\ G_n \\ [a,r] \end{bmatrix} = a \left[\frac{\frac{(r^{(n+1)} - 1)}{(r - 1)} - n}{(r - 1)} \right] = a \left[\frac{r^{(n+1)} - 1 - n(r - 1)}{(r - 1)^2} \right]$$

$$= a \left[\frac{r^{(n+1)} - [n(r - 1) + 1]}{(r - 1)^2} \right]$$

$$= a \left[\frac{r^{(n+1)} - \frac{[n^2(r - 1)^2 - 1]}{[n - (r - 1) - 1]}}{(r - 1)^2} \right]$$

$$= a \left[\frac{r^{(n+1)}[n(r - 1) - 1] - [n^2(r - 1)^2 - 1]}{(r - 1)^2[n(r - 1) - 1]} \right] = \begin{matrix} [3] \\ G_n \\ [a,r] \end{matrix}$$

$$[S]_n \begin{bmatrix} [3] \\ G_n \\ [a,r] \end{bmatrix} = a \left[\frac{r^{(n+2)} - [n^2(r - 1)^2 + n(r - 1) + 1]}{(r - 1)^3} \right]$$

$$= a \left[\frac{r^{(n+1)} - \frac{[n^3(r - 1)^3 - 1]}{[n - (r - 1) - 1]}}{(r - 1)^3} \right]$$

$$= a \left[\frac{r^{(n+1)}[n - (r - 1) - 1] - [n^3(r - 1)^3 - 1]}{(r - 1)^3[n(r - 1) - 1]} \right] = \begin{matrix} [4] \\ G_n \\ [a,r] \end{matrix}$$

$$[S]_n \begin{bmatrix} [4] \\ G_n \\ [a,r] \end{bmatrix} = a \left[\frac{r^{(n+3)} - [n^3(r-1)^3 + n^2(r-1)^2 + n(r-1) + 1]}{(r-1)^4} \right]$$

$$= a \left[\frac{r^{(n+1)} - \dfrac{[n^4(r-1)^4 - 1]}{[n - (r-1) - 1]}}{(r-1)^4} \right]$$

$$= a \left[\frac{r^{(n+1)}[n(r-1) - 1] - [n^4(r-1)^4 - 1]}{(r-1)^4[n(r-1) - 1]} \right] = \begin{matrix} [5] \\ G_n \\ [a,r] \end{matrix}$$

$$[S]_n \begin{bmatrix} [5] \\ G_n \\ [a,r] \end{bmatrix} = a \left[\frac{r^{(n+4)} - [n^4(r-1)^4 + n^3(r-1)^3 + n^2(r-1)^2 + n(r-1) + 1]}{(r-1)^5} \right]$$

$$= a \left[\frac{r^{(n+1)} - \dfrac{[n^5(r-1)^5 - 1]}{[n(r-1) - 1]}}{(r-1)^5} \right]$$

$$= a \left[\frac{r^{(n+1)}[n(r-1) - 1] - [n^5(r-1)^5 - 1]}{(r-1)^5[n(r-1) - 1]} \right] = \begin{matrix} [6] \\ G_n \\ [a,r] \end{matrix}$$

$$-------------\ggg\ggg\ggg\ggg$$

$$[S]_n \begin{bmatrix} [\alpha - 1] \\ G_n \\ [a,r] \end{bmatrix}$$

$$= a \left[\frac{r^{(n+\alpha-2)} - [n^{(\alpha-2)}(r-1)^{(\alpha-2)} + \cdots + n^2(r-1)^2 + n(r-1) + 1]}{(r-1)^{(\alpha-1)}} \right]$$

$$= a \left[\frac{r^{(n+1)} - \dfrac{[n^{(\alpha-1)}(r-1)^{(\alpha-1)} - 1]}{[n(r-1) - 1]}}{(r-1)^{(\alpha-1)}} \right]$$

$$= a \left[\frac{r^{(n+1)}[n(r-1) - 1] - [n^{(\alpha-1)}(r-1)^{(\alpha-1)} - 1]}{(r-1)^{(\alpha-1)}[n(r-1) - 1]} \right] = \begin{matrix} [\alpha] \\ G_n \\ [a,r] \end{matrix}$$

$$[S]_n \begin{bmatrix} [\alpha] \\ G_n \\ [a,r] \end{bmatrix}$$

$$= a \left[\frac{r^{(n+\alpha-1)} - [n^{(\alpha-1)}(r-1)^{(\alpha-1)} + \cdots + n^2(r-1)^2 + n(r-1) + 1]}{(r-1)^\alpha} \right]$$

$$= a \left[\frac{r^{(n+1)} - \dfrac{[n^\alpha(r-1)^\alpha - 1]}{[n(r-1)-1]}}{(r-1)^\alpha} \right]$$

$$= a \left[\frac{r^{(n+1)}[n(r-1)-1] - [n^\alpha(r-1)^\alpha - 1]}{(r-1)^\alpha[n(r-1)-1]} \right] = \begin{matrix} [\alpha+1] \\ G_n \\ [a,r] \end{matrix}$$

$$[S]_n \begin{bmatrix} [\alpha+1] \\ G_n \\ [a,r] \end{bmatrix}$$

$$= a \left[\frac{r^{(n+\alpha)} - [n^\alpha(r-1)^\alpha + \cdots + n^2(r-1)^2 + n(r-1) + 1]}{(r-1)^{(\alpha+1)}} \right]$$

$$= a \left[\frac{r^{(n+1)} - \dfrac{[n^{(\alpha+1)}(r-1)^{(\alpha+1)} - 1]}{[n(r-1)-1]}}{(r-1)^{(\alpha+1)}} \right]$$

$$= a \left[\frac{r^{(n+1)}[n(r-1)-1] - [n^{(\alpha+1)}(r-1)^{(\alpha+1)} - 1]}{(r-1)^{(\alpha+1)}[n(r-1)-1]} \right] = \begin{matrix} [\alpha+2] \\ G_n \\ [a,r] \end{matrix}$$

-------------------->>>>>>>>>>>>>>>>>>> ∞∞ FOR EVER ∞

2. Inductive Geometric Progressions

We define the INDUCTIVE–GEOMETRIC PROGRESSION

$$\begin{bmatrix} GI_n \\ [a,r,s] \end{bmatrix} = ar^{(n-1)}s^{\left(\frac{(n-2)(n-1)}{2!}\right)} = \begin{matrix} [1] \\ GI_n \\ [a,r,s] \end{matrix} \qquad [n \geq 1]$$

"a" as the Initiating Term and "r" as the Common Ratio and "s" is the Inductive Ratio [a, r, and s are algebraic numbers ($r \neq 1$]. We can determine the Super–Sums upto any Level $\alpha = 1, 2, 3 \cdots \infty$. The ordinary Sum of the INDUCTIVE–GEOMETRIC PROGRESSION is clearly the Super–Sum Level -1.

Notation 2.1. In $[S]_n \begin{bmatrix} [\alpha] \\ GI_n \\ [a,r,s] \end{bmatrix}$, $[S]_n$ is the Sum upto n terms of the αth Level of the INDUCTIVE–GEOMETRIC PROGRESSION [Super–Sum α^{th} Level] and it yields the $\begin{bmatrix} [\alpha+1] \\ GI_n \\ [a,r,s] \end{bmatrix}$ the $(\alpha+1)^{th}$ Level of the given INDUCTIVE– GEOMETRIC PROGRESSION

$$\begin{bmatrix} GI_n \\ [a,r,s] \end{bmatrix} = ar^{(n-1)}s^{\left(\frac{(n-2)(n-1)}{2!}\right)} = \begin{matrix} [1] \\ GI_n \\ [a,r,s] \end{matrix}$$

ALL the following Sequence of Formulae could be easily Proved by the Method of Mathematical Induction by now traditionally formalized. For each Level, the induction is performed on "n" and for The General Result the induction is performed on "α". The Routine Steps are omitted to save Eternal Space–Time!

$$[S]_n \begin{bmatrix} [1] \\ GI_n \\ [a,r,s] \end{bmatrix} = \frac{a(r^n-1)}{(r-1)} \left[\sum_{i=2}^{(n-1)} (n-i)s^{\left(\frac{(n-i)(n-i+1)}{2!}\right)} \right]$$

$$= \begin{matrix} [2] \\ GI_n \\ [a,r,s] \end{matrix}$$

$$[S]_n \begin{bmatrix} [1] \\ GI_n \\ [a,r,s] \end{bmatrix} = a \left[\frac{\frac{(r^{(n+1)}-1)}{(r-1)}-n}{(r-1)} \right]$$

$$\left[\sum_{i=2}^{(n-1)} \left(\frac{(n-i)(n-i+1)}{2!}\right)s^{\left(\frac{(n-i)(n-i+1)}{2!}\right)} \right]$$

$$= a \left[\frac{r^{(n+1)}-1-n(r-1)}{(r-1)^2} \right]$$

$$\left[\sum_{i=2}^{(n-1)} \left(\frac{(n-i)(n-i+1)}{2!}\right)s^{\left(\frac{(n-i)(n-i+1)}{2!}\right)} \right]$$

$$= a \left[\frac{r^{(n+1)} - [n(r-1)+1]}{(r-1)^2} \right] \left[\sum_{i=2}^{(n-1)} \left(\frac{(n-i)(n-i+1)}{2!} \right) s^{\left(\frac{(n-i)(n-i+1)}{2!} \right)} \right]$$

$$= a \left[\frac{r^{(n+1)} - \frac{[n^2(r-1)^2-1]}{[n(r-1)-1]}}{(r-1)^2} \right] \left[\sum_{i=2}^{(n-1)} \left(\frac{(n-i)(n-i+1)}{2!} \right) s^{\left(\frac{(n-i)(n-i+1)}{2!} \right)} \right]$$

$$= a \left[\frac{r^{(n+1)}[n(r-1)-1] - [n^2(r-1)^2-1]}{(r-1)^2[n(r-1)-1]} \right]$$

$$\left[\sum_{i=2}^{(n-1)} \left(\frac{(n-i)(n-i+1)}{2!} \right) s^{\left(\frac{(n-i)(n-i+1)}{2!} \right)} \right] = \begin{array}{c} [3] \\ GI_n \\ [a,r,s] \end{array}$$

$$[S]_n \begin{bmatrix} [3] \\ GI_n \\ [a,r,s] \end{bmatrix} = a \left[\frac{r^{(n+2)} - [n^2(r-1)^2 + n(r-1)+1]}{(r-1)^3} \right]$$

$$\left[\sum_{i=2}^{(n-1)} \left(\frac{(n-i)(n-i+1)(n-i+2)}{3!} \right) s^{\left(\frac{(n-i)(n-i+1)}{2!} \right)} \right]$$

$$= a \left[\frac{r^{(n+1)} - \frac{[n^3(r-1)^3-1]}{[n(r-1)-1]}}{(r-1)^3} \right]$$

$$\left[\sum_{i=2}^{(n-1)} \left(\frac{(n-i)(n-i+1)(n-i+2)}{3!} \right) s^{\left(\frac{(n-i)(n-i+1)}{2!} \right)} \right]$$

$$= a \left[\frac{r^{(n+1)}[n(r-1)-1] - [n^3(r-1)^3-1]}{(r-1)^3[n(r-1)-1]} \right]$$

$$\left[\sum_{i=2}^{(n-1)} \left(\frac{(n-i)(n-i+1)(n-i+2)}{3!} \right) s^{\left(\frac{(n-i)(n-i+1)}{2!} \right)} \right]$$

$$= \begin{array}{c} [4] \\ GI_n \\ [a,r,s] \end{array}$$

$$[S]_n \begin{bmatrix} [4] \\ GI_n \\ [a,r,s] \end{bmatrix} = a \left[\frac{r^{(n+3)} - [n^3(r-1)^3 + n^2(r-1)^2 + n(r-1) + 1]}{(r-1)^4} \right]$$

$$\left[\sum_{i=2}^{(n-1)} \left(\frac{(n-i)(n-i+1)(n-i+2)(n-i+3)}{4!} \right) {}_s\left(\frac{(n-i)(n-i+1)}{2!} \right) \right]$$

$$= a \left[\frac{r^{(n+1)} - \dfrac{[n^4(r-1)^4 - 1]}{[n(r-1)-1]}}{(r-1)^4} \right]$$

$$\left[\sum_{i=2}^{(n-1)} \left(\frac{(n-i)(n-i+1)(n-i+2)(n-i+3)}{4!} \right) {}_s\left(\frac{(n-i)(n-i+1)}{2!} \right) \right]$$

$$= a \left[\frac{r^{(n+1)}[n(r-1)-1] - [n^4(r-1)^4 - 1]}{(r-1)^4[n(r-1)-1]} \right]$$

$$\left[\sum_{i=2}^{(n-1)} \left(\frac{(n-i)(n-i+1)(n-i+2)(n-i+3)}{4!} \right) {}_s\left(\frac{(n-i)(n-i+1)}{2!} \right) \right]$$

$$= \begin{matrix} [5] \\ GI_n \\ [a,r,s] \end{matrix}$$

$$[S]_n \begin{bmatrix} [5] \\ GI_n \\ [a,r,s] \end{bmatrix}$$

$$= a \left[\frac{r^{(n+4)} - [n^4(r-1)^4 + n^3(r-1)^3 + n^2(r-1)^2 + n(r-1) + 1]}{(r-1)^5} \right]$$

$$\left[\sum_{i=2}^{(n-1)} \left(\frac{(n-i)(n-i+1)(n-i+2)(n-i+3)(n-i+4)}{5!} \right) \right.$$

$$\left. {}_s\left(\frac{(n-i)(n-i+1)}{2!} \right) \right]$$

$$= a \left[\frac{r^{\cdot(n+1)} - \dfrac{[n^5(r-1)^5 - 1]}{[n(r-1) - 1]}}{(r-1)^5} \right]$$

$$\left[\sum_{i=2}^{(n-1)} \left(\frac{(n-i)(n-i+1)(n-i+2)(n-i+3)(n-i+4)}{5!} \right) \right.$$

$$\left. {}_s\left(\frac{(n-i)(n-i+1)}{2!} \right) \right]$$

$$= a \left[\frac{r^{(n+1)}[n(r-1) - 1] - [n^5(r-1)^5 - 1]}{(r-1)^5[n(r-1) - 1]} \right]$$

$$\left[\sum_{i=2}^{(n-1)} \left(\frac{(n-i)(n-i+1)(n-i+2)(n-i+3)(n-i+4)}{5!} \right) \right.$$

$$\left. {}_s\left(\frac{(n-i)(n-i+1)}{2!} \right) \right] = \begin{matrix} [6] \\ GI_n \\ [a,r,s] \end{matrix}$$

- - - - - - - - - - - - ->》》》》》》》》》》》》》》》》》》》》》》》》》》》》》》》》》》》》

$$[S]_n \begin{bmatrix} [\alpha - 1] \\ GI_n \\ [a,r,s] \end{bmatrix}$$

$$= a \left[\frac{r^{(n+\alpha-2)} - [n^{(\alpha-2)}(r-1)^{(\alpha-2)} + \cdots + n^2(r-1)^2 + n(r-1) + 1]}{(r-1)^{(\alpha-1)}} \right]$$

$$\left[\sum_{i=2}^{(n-1)} \left(\frac{(n-i)(n-i+1)\cdots(n-i+\alpha-2)}{(\alpha-1)!} \right) {}_s\left(\frac{(n-i)(n-i+1)}{2!} \right) \right]$$

$$= a \left[\frac{r^{\cdot(n+1)} - \dfrac{[n^{(\alpha-1)}(r-1)^{(\alpha-1)} - 1]}{[n(r-1) - 1]}}{(r-1)^{(\alpha-1)}} \right]$$

$$\left[\sum_{i=2}^{(n-1)} \left(\frac{(n-i)(n-i+1)\cdots(n-i+\alpha-2)}{(\alpha-1)!} \right) {}_s\left(\frac{(n-i)(n-i+1)}{2!} \right) \right]$$

$$= a \left[\frac{r^{(n+1)}[n(r-1)-1] - [n^{(\alpha-1)}(r-1)^{(\alpha-1)} - 1]}{(r-1)^{(\alpha-1)}[n(r-1)-1]} \right]$$

$$\left[\sum_{i=2}^{(n-1)} \left(\frac{(n-i)(n-i+1)\cdots(n-i+\alpha-2)}{(\alpha-1)!} \right) s \left(\frac{(n-i)(n-i+1)}{2!} \right) \right]$$

$$= \begin{matrix} [\alpha] \\ GI_n \\ [a,r,s] \end{matrix}$$

$$[S]_n \begin{bmatrix} [\alpha] \\ GI_n \\ [a,r,s] \end{bmatrix}$$

$$= a \left[\frac{r^{(n+\alpha-1)} - [n^{(\alpha-1)}(r-1)^{(\alpha-1)} + \cdots + n^2(r-1)^2 + n(r-1) + 1]}{(r-1)^\alpha} \right]$$

$$\left[\sum_{i=2}^{(n-1)} \left(\frac{(n-i)(n-i+1)\cdots(n-i+\alpha-1)}{\alpha!} \right) s \left(\frac{(n-i)(n-i+1)}{2!} \right) \right]$$

$$= a \left[\frac{r^{(n+1)} - \dfrac{[n^\alpha(r-1)^\alpha - 1]}{[n(r-1)-1]}}{(r-1)^\alpha} \right]$$

$$\left[\sum_{i=2}^{(n-1)} \left(\frac{(n-i)(n-i+1)\cdots(n-i+\alpha-1)}{\alpha!} \right) s \left(\frac{(n-i)(n-i+1)}{2!} \right) \right]$$

$$= a \left[\frac{r^{(n+1)}[n(r-1)-1] - [n^\alpha(r-1)^\alpha - 1]}{(r-1)^\alpha[n(r-1)-1]} \right]$$

$$\left[\sum_{i=2}^{(n-1)} \left(\frac{(n-i)(n-i+1)\cdots(n-i+\alpha-1)}{\alpha!} \right) s \left(\frac{(n-i)(n-i+1)}{2!} \right) \right]$$

$$= \begin{matrix} [\alpha+1] \\ GI_n \\ [a,r,s] \end{matrix}$$

$$[S]_n \begin{bmatrix} [\alpha + 1] \\ GI_n \\ [a, r, s] \end{bmatrix}$$

$$= a \left[\frac{r^{(n+\alpha)} - [n^\alpha (r-1)^\alpha + \cdots + n^2 (r-1)^2 + n(r-1) + 1]}{(r-1)^{(\alpha+1)}} \right]$$

$$\left[\sum_{i=2}^{(n-1)} \left(\frac{(n-i)(n-i+1) \cdots (n-i+\alpha)}{(\alpha+1)!} \right)_s \left(\frac{(n-i)(n-i+1)}{2!} \right) \right]$$

$$= a \left[\frac{r^{(n+1)} - \frac{[n^{(\alpha+1)}(r-1)^{(\alpha+1)} - 1]}{[n(r-1) - 1]}}{(r-1)^{(\alpha+1)}} \right]$$

$$\left[\sum_{i=2}^{(n-1)} \left(\frac{(n-i)(n-i+1) \cdots (n-i+\alpha)}{(\alpha+1)!} \right)_s \left(\frac{(n-i)(n-i+1)}{2!} \right) \right]$$

$$= a \left[\frac{r^{(n+1)}[n(r-1) - 1] - [n^{(\alpha+1)}(r-1)^{(\alpha+1)} - 1]}{(r-1)^{(\alpha+1)}[n(r-1) - 1]} \right]$$

$$\left[\sum_{i=2}^{(n-1)} \left(\frac{(n-i)(n-i+1) \cdots (n-i+\alpha)}{(\alpha+1)!} \right)_s \left(\frac{(n-i)(n-i+1)}{2!} \right) \right]$$

$$= \begin{matrix} [\alpha + 2] \\ GI_n \\ [a, r, s] \end{matrix}$$

- - - - - - - - - - - - - - - ->≫≫≫≫≫≫≫≫≫≫≫ ∞∞ FOR EVER ∞

3. Generalized Geometric Progressions–Simple [$*$ Geometric Progressions]

We define the GENERALIZED GEOMETRIC PROGRESSION–SIMPLE [$*$ GEOMETRIC PROGRESSION]

$$\left[\begin{matrix} G_n^* \\ [a,r,s_1,\text{---}s_\beta] \end{matrix}\right] = ar^{(n-1)}s_1^{\left(\frac{(n-2)(n-1)}{2!}\right)} \sum_{i=3}^{(n-1)} (n-i)s_2^{\left(\frac{(n-i)(n-i+1)}{2!}\right)}$$

$$\sum_{i=4}^{(n-1)} \left(\frac{(n-i)(n-i+1)}{2!}\right)s_3^{\left(\frac{(n-i)(n-i+1)}{2!}\right)}\ldots,$$

$$\sum_{i=\beta+1}^{(n-1)} \left(\frac{(n-i)(n-i+1)\text{---}(n-i+\beta-2)}{(\beta-1)!}\right)$$

$$s_\beta^{\left(\frac{(n-i)(n-i+1)}{2!}\right)}$$

$$= ar^{(n-1)} \sum_{\phi=1}^{\beta} \sum_{i=\phi+1}^{(n-1)} \left(\frac{(n-i)(n-i+1)\text{---}(n-i+\phi-2)}{(\phi-1)!}\right)$$

$$s_\phi^{\left(\frac{(n-i)(n-i+1)}{2!}\right)}$$

$$= \begin{matrix} [1] \\ G_n^* \\ [a,r,s_1,\text{---},s_\beta] \end{matrix} \qquad [n \geq 1]$$

"a" as the Initiating Term, "r" as the Common Ratio and "s_1", "s_2", "s_3", ---, "s_β" as the Inductive Ratios [$a,r,s_1,s_2,s_3,\text{---},s_\beta$ are algebraic numbers]. We can determine the Super–Sums of the GENERALIZED GEOMETRIC PROGRESSION–SIMPLE upto any Level $\alpha = 1,2,3\text{---}\infty$. The ordinary Sum of GENERALIZED GEOMETRIC PROGRESSION – SIMPLE is clearly the Super–Sum Level -1 of the same "β" may be called Generalization–Level of the GENERALIZED GEOMETRIC PROGRESSION –SIMPLE.

Notation 3.1. In $[S]_n \begin{bmatrix} [\alpha] \\ G_n^* \\ [a,r,s_1,\text{---},s_\beta] \end{bmatrix}$, $[S]_n$ is the Sum upto n terms of the

α^{th} Level of the GENERALIZED GEOMETRIC PROGRESSION –SIMPLE

[Super–Sum α^{th} Level] and it yields the $\begin{matrix} [\alpha+1] \\ G_n^* \\ [a,r,s_1,\cdots,s_\beta] \end{matrix}$ the $(\alpha+1)^{th}$ Level of

the given GENERALIZED GEOMETRIC PROGRESSION –SIMPLE.

$$\begin{bmatrix} G_n^* \\ [a,r,s_1,\text{---},s_\beta] \end{bmatrix} = ar^{(n-1)} \sum_{\phi=1}^{\beta} \sum_{i=\phi+1}^{(n-1)} \left(\frac{(n-i)(n-i+1)\cdots(n-i+\phi-2)}{(\phi-1)!} \right)$$

$$\left(\frac{(n-i)(n-i+1)}{2!} \right)_{s_\phi}$$

$$= \begin{matrix} [1] \\ G_n^* \\ [a,r,s_1,\text{---},s_\beta] \end{matrix}$$

ALL the following Sequence of Formulae could be easily proved by the method of Mathematical induction by now traditionally formalized. For each Level, the induction is performed on "n" and for The General Result the induction is performed on "α". The Routine Steps are omitted to save Eternal Space–Time!

$$[S]_n \begin{bmatrix} [1] \\ G_n^* \\ [a,r,s_1,\text{---},s_\beta] \end{bmatrix} = \frac{a(r^n-1)}{(r-1)} \sum_{\phi=1}^{\beta} \sum_{i=\phi+1}^{(n-1)}$$

$$\left(\frac{(n-i)(n-i+1)\text{---}(n-i+\phi-1)}{\phi!} \right)$$

$$\left(\frac{(n-i)(n-i+1)}{2!} \right)_{s_\phi}$$

$$= \begin{bmatrix} [2] \\ G_n^* \\ [a,r,s_1,\text{---},s_\beta] \end{bmatrix}$$

$$[S]_n \begin{bmatrix} \overset{[2]}{G_n^*} \\ [a, r, s_1, \cdots, s_\beta] \end{bmatrix} = a \left[\frac{r^{(n+1)}[n(r-1)-1] - [n^2(r-1)^2 - 1]}{(r-1)^2[n(r-1)-1]} \right]$$

$$\sum_{\phi=1}^{\beta} \sum_{i=\phi+1}^{(n-1)} \left(\frac{(n-i)(n-i+1)\cdots(n-i+\phi)}{(\phi+1)!} \right)$$

$$s_\phi \left(\frac{(n-i)(n-i+1)}{2!} \right)$$

$$= \underset{[a, r, s_1, \cdots, s_\beta]}{\overset{[3]}{G_n^*}}$$

$$[S]_n \begin{bmatrix} \overset{[3]}{G_n^*} \\ [a, r, s_1, \cdots, s_\beta] \end{bmatrix} = a \left[\frac{r^{(n+1)}[n(r-1)-1] - [n^3(r-1)^3 - 1]}{(r-1)^3[n(r-1)-1]} \right]$$

$$\sum_{\phi=1}^{\beta} \sum_{i=\phi+1}^{(n-1)} \left(\frac{(n-i)(n-i+1)\text{-}\cdots\text{-}(n-i+\phi+1)}{(\phi+2)!} \right)$$

$$s_\phi \left(\frac{(n-i)(n-i+1)}{2!} \right)$$

$$= \underset{[a, r, s_1, \cdots, s_\beta]}{\overset{[4]}{G_n^*}}$$

$$[S]_n \begin{bmatrix} \overset{[4]}{G_n^*} \\ [a, r, s_1, \cdots, s_\beta] \end{bmatrix} = a \left[\frac{r^{(n+1)}[n(r-1)-1] - [n^4(r-1)^4 - 1]}{(r-1)^4[n(r-1)-1]} \right]$$

$$\sum_{\phi=1}^{\beta} \sum_{i=\phi+1}^{(n-1)} \left(\frac{(n-i)(n-i+1)\cdots(n-i+\phi+2)}{(\phi+3)!} \right)$$

$$s_\phi \left(\frac{(n-i)(n-i+1)}{2!} \right)$$

$$= \underset{[a, r, s_1, \text{---}, s_\beta]}{\overset{[5]}{G_n^*}}$$

$$[S]_n \begin{bmatrix} [5] \\ G_n^* \\ [a, r, s_1, \cdots, s_\beta] \end{bmatrix} = a \left[\frac{r^{(n+1)}[n(r-1)-1] - [n^5(r-1)^5 - 1]}{(r-1)^5[n(r-1)-1]} \right]$$

$$\sum_{\phi=1}^{\beta} \sum_{i=\phi+1}^{(n-1)} \left(\frac{(n-i)(n-i+1)\cdots(n-i+\phi+3)}{(\phi+4)!} \right)$$

$$s_\phi \left(\frac{(n-i)(n-i+1)}{2!} \right)$$

$$= \begin{array}{c} [6] \\ G_n^* \\ [a, r, s_1, \cdots, s_\beta] \end{array}$$

- - - - - - - - - - - - - - - →≫≫≫≫≫≫≫≫≫≫≫≫≫≫≫≫≫≫≫≫≫

$$[S]_n \begin{bmatrix} [\alpha - 1] \\ G_n^* \\ [a, r, s_1, \cdots, s_\beta] \end{bmatrix} = a \left[\frac{r^{(n+1)}[n(r-1)-1] - [n^{(\alpha-1)}(r-1)^{(\alpha-1)} - 1]}{(r-1)^{(\alpha-1)}[n(r-1)-1]} \right]$$

$$\sum_{\phi=1}^{\beta} \sum_{i=\phi+1}^{(n-1)} \left(\frac{(n-i)(n-i+1)\cdots(n-i+\phi+\alpha-3)}{(\phi+\alpha-2)!} \right)$$

$$s_\phi \left(\frac{(n-i)(n-i+1)}{2!} \right)$$

$$= \begin{array}{c} [\alpha] \\ G_n^* \\ [a, r, s_1, ---, s_\beta] \end{array}$$

$$[S]_n \begin{bmatrix} [\alpha] \\ G_n^* \\ [a, r, s_1, \cdots, s_\beta] \end{bmatrix} = a \left[\frac{r^{(n+1)}[n(r-1)-1] - [n^\alpha(r-1)^\alpha - 1]}{(r-1)^\alpha[n(r-1)-1]} \right]$$

$$\sum_{\phi=1}^{\beta} \sum_{i=\phi+1}^{(n-1)} \left(\frac{(n-i)(n-i+1)\cdots(n-i+\phi+\alpha-2)}{(\phi+\alpha-1)!} \right)$$

$$s_\phi \left(\frac{(n-i)(n-i+1)}{2!} \right)$$

$$= \begin{array}{c} [\alpha + 1] \\ G_n^* \\ [a, r, s_1, ---, s_\beta] \end{array}$$

$$[S]_n \begin{bmatrix} [\alpha+1] \\ G_n^* \\ [a,r,s_1,\cdots,s_\beta] \end{bmatrix} = a \left[\frac{r^{(n+1)}[n(r-1)-1] - [n^{(\alpha+1)}(r-1)^{(\alpha+1)}-1]}{(r-1)^{(\alpha+1)}[n(r-1)-1]} \right]$$

$$\sum_{\phi=1}^{\beta} \sum_{i=\phi+1}^{(n-1)} \left(\frac{(n-i)(n-i+1)\cdots(n-i+\phi+\alpha-1)}{(\phi+\alpha)!} \right)$$

$$s_\phi \left(\frac{(n-i)(n-i+1)}{2!} \right)$$

$$= \begin{matrix} [\alpha+2] \\ G_n^* \\ [a,r,s_1,\cdots,s_\beta] \end{matrix}$$

- - - - - - - - - - - - - - ->≫≫≫≫≫≫≫≫≫≫ ∞∞ FOR EVER ∞

4. Generalized Geometric Progressions–Random [$*^r$ Geometric Progressions]

We define the GENERALIZED GEOMETRIC PROGRESSIONS–RANDOM [$*^r$ GEOMETRIC PROGRESSION]

$$\begin{bmatrix} G_n^{*^r} \\ [a,r,s_1,\cdots,s_\beta] \\ [@,2,j_1,\cdots,j_\beta] \end{bmatrix} = ar^{(n-1)}s_1 \left(\frac{(n-j_1)(n-j_1+1)}{2!} \right)$$

$$\sum_{i=j_2}^{(n-1)} (n-i)s_2 \left(\frac{(n-i)(n-i+1)}{2!} \right)$$

$$\sum_{i=j_3}^{(n-1)} \left(\frac{(n-i)(n-i+1)}{2!} \right) s_3 \left(\frac{(n-i)(n-i+1)}{2!} \right) \cdots$$

$$\sum_{i=j_\beta}^{(n-1)} \left(\frac{(n-i)(n-i+1)\cdots(n-i+\beta-2)}{(\beta-1)!} \right)$$

$$s_\beta \left(\frac{(n-i)(n-i+1)}{2!} \right)$$

$$= ar^{(n-1)} \sum_{\phi=1}^{\beta} \sum_{i=j_\phi}^{(n-1)} \left(\frac{(n-i)(n-i+1)\cdots(n-i+\phi-2)}{(\phi-1)!} \right) s_\phi^{\left(\frac{(n-i)(n-i+1)}{2!} \right)}$$

$$= \underset{\substack{[a,r,s_1,\cdots,s_\beta] \\ [@,2,j_1,\cdots,j_\beta]}}{\overset{[1]}{G_n^{*}}^{r}} \qquad [n \geq 1]$$

"a" as the initiating Term, "r" as the Common Ratio and "s_1", "s_2", "s_3", - - - , "s_β" as the inductive Ratios [$a, r, s_1, s_2, s_3, - - -, s_\beta$ are algebraic number $r \neq 1$] induced at [@] "j_1", "j_2", "j_3", - - - , "j_β" respectively [$j_1 > 1, i = 1, 2 - - - \beta$].

We can determine the Super–Sums of the GENERALIZED GEOMETRIC PROGRESSION– RANDOM up to any Level $\alpha = 1, 2, 3 - - - \infty$. The ordinary Sum of GENERALIZED GEOMETRIC PROGRESSION– RANDOM is clearly the Super–Sum Level -1 of the same. "β" may be called Generalization–Level of the GENERALIZED GEOMETRIC PROGRESSION –RANDOM.

Notation 4.1. In $[S]_n \begin{bmatrix} [\alpha] \\ G_n^{*}{}^{r} \\ [a,r,s_1,\cdots,s_\beta] \\ [@,s,j_1,\cdots,j_\beta] \end{bmatrix}$, $[S]_n$ is the Sum upto n terms of the αth Level of the GENERALIZED GEOMETRIC PROGRESSION – RANDOM

[Super–Sum α^{th} Level] and it yields the $\underset{\substack{[a,r,s_1,\cdots,s_\beta] \\ [@,2,j_1,\cdots,j_\beta]}}{\overset{[\alpha+1]}{G_n^{*}}^{r}}$ the $(\alpha+1)^{th}$ Level of the given Generalized Geometric Progression–Random.

$$\begin{bmatrix} G_n^{*}{}^{r} \\ [a,r,s_1,\cdots,s_\beta] \\ [@,2,j_1,\cdots,j_\beta] \end{bmatrix} = ar^{(n-1)} \sum_{\phi=1}^{\beta} \sum_{i=j_\phi}^{(n-1)} \left(\frac{(n-i)(n-i+1)\cdots(n-i+\phi-2)}{(\phi-1)!} \right)$$

$$s_\phi^{\left(\frac{(n-i)(n-i+1)}{2!} \right)}$$

$$= \underset{\substack{[a,r,s_1,---,s_\beta] \\ [@,2,j_1,---,j_\beta]}}{\overset{[1]}{G_n^{*}}^{r}}$$

ALL the following Sequence of Formulae could be easily Proved by the
Method of Mathematical Induction by now traditionally formalized. For each
Level, the induction is performed on "n" and for The General Result the in-
duction is performed on "α". The Routine Steps are omitted to save Eternal
Space–Time !

$$[S]_n \begin{bmatrix} [1] \\ G_n^{*^r} \\ {[a, r, s_1, \cdots, s_\beta]} \\ {[@, 2, j_1, \cdots, j_\beta]} \end{bmatrix} = \frac{a(r^n - 1)}{(r-1)} \sum_{\phi=1}^{\beta} \sum_{i=j_\phi}^{(n-1)}$$

$$\left(\frac{(n-i)(n-i+1)\cdots(n-i+\phi-1)}{\phi!} \right)$$

$$_{s_\phi}\left(\frac{(n-i)(n-i+1)}{2!} \right)$$

$$= \begin{matrix} [2] \\ G_n^{*^r} \\ {[a, r, s_1, \cdots, s_\beta]} \\ {[@, 2, j_1, \cdots, j_\beta]} \end{matrix}$$

$$[S]_n \begin{bmatrix} [2] \\ G_n^{*^r} \\ {[a, r, s_1, \cdots, s_\beta]} \\ {[@, 2, j_1, \cdots, j_\beta]} \end{bmatrix} = a \left[\frac{r^{2(n+1)}[n(r-1)-1] - [n^2(r-1)^2 - 1]}{(r-1)^2[n(r-1)-1]} \right]$$

$$\sum_{\phi=1}^{\beta} \sum_{i=j_\phi}^{(n-1)} \left(\frac{(n-i)(n-i+1)\cdots(n-i+\phi)}{(\phi+1)!} \right)$$

$$_{s_\phi}\left(\frac{(n-i)(n-i+1)}{2!} \right)$$

$$= \begin{matrix} [3] \\ G_n^{*^r} \\ {[a, r, s_1, \cdots, s_\beta]} \\ {[@, 2, j_1, \cdots, j_\beta]} \end{matrix}$$

$$[S]_n \begin{bmatrix} [3] \\ G_n^{*^r} \\ {[a, r, s_1, \cdots, s_\beta]} \\ {[@, 2, j_1, \cdots, j_\beta]} \end{bmatrix} = a \left[\frac{r^{(n+1)}[n(r-1)-1] - [n^3(r-1)^3 - 1]}{(r-1)^3[n(r-1)-1]} \right]$$

$$\sum_{\phi=1}^{\beta} \sum_{i=j_\phi}^{(n-1)} \left(\frac{(n-i)(n-i+1)\cdots(n-i+\phi+1)}{(\phi+2)!} \right) s_\phi \left(\frac{(n-i)(n-i+1)}{2!} \right)$$

$$= \quad \begin{matrix} [4] \\ G^{*\,r}_{\,n} \\ [a,r,s_1,\cdots,s_\beta] \\ [@,2,j_1,\cdots,j_\beta] \end{matrix}$$

$$[S]_n \begin{bmatrix} [4] \\ G^{*\,r}_{\,n} \\ [a,r,s_1,\cdots,s_\beta] \\ [@,2,j_1,\cdots,j_\beta] \end{bmatrix} = a \left[\frac{r^{(n+1)}[n(r-1)-1] - [n^4(r-1)^4 - 1]}{(r-1)^4[n(r-1)-1]} \right]$$

$$\sum_{\phi=1}^{\beta} \sum_{i=j_\phi}^{(n-1)} \left(\frac{(n-i)(n-i+1)\cdots(n-i+\phi+2)}{(\phi+3)!} \right)$$

$$s_\phi \left(\frac{(n-i)(n-i+1)}{2!} \right)$$

$$= \quad \begin{matrix} [5] \\ G^{*\,r}_{\,n} \\ [a,r,s_1,\cdots,s_\beta] \\ [@,2,j_1,\cdots,j_\beta] \end{matrix}$$

$$S]_n \begin{bmatrix} [5] \\ G^{*\,r}_{\,n} \\ [a,r,s_1,\cdots,s_\beta] \\ [@,2,j_1,\cdots,j_\beta] \end{bmatrix} = a \left[\frac{r^{(n+1)}[n(r-1)-1] - [n^5(r-1)^5 - 1]}{(r-1)^5[n(r-1)-1]} \right]$$

$$\sum_{\phi=1}^{\beta} \sum_{i=j_\phi}^{(n-1)} \left(\frac{(n-i)(n-i+1)\cdots(n-i+\phi+3)}{(\phi+4)!} \right)$$

$$s_\phi \left(\frac{(n-i)(n-i+1)}{2!} \right)$$

$$= \quad \begin{matrix} [6] \\ G^{*\,r}_{\,n} \\ [a,r,s_1,\cdots,s_\beta] \\ [@,2,j_1,\cdots,j_\beta] \end{matrix}$$

- - - - - - - - - - - - - - - ->≫≫≫≫≫≫≫≫≫≫≫≫≫≫≫≫≫≫≫≫

$$[S]_n \begin{bmatrix} [\alpha-1] \\ G_n^{*r} \\ [a,r,s_1,\cdots,s_\beta] \\ [@,2,j_1,\cdots,j_\beta] \end{bmatrix} = a \left[\frac{r^{(n+1)}[n(r-1)-1] - [n^{(\alpha-1)}(r-1)^{(\alpha-1)} - 1]}{(r-1)^{(\alpha-1)}[n(r-1)-1]} \right]$$

$$\sum_{\phi=1}^{\beta} \sum_{i=j_\phi}^{(n-1)} \left(\frac{(n-i)(n-i+1)\cdots(n-i+\phi+\alpha-3)}{(\phi+\alpha-2)!} \right) s_\phi^{\left(\frac{(n-i)(n-i+1)}{2!} \right)}$$

$$= \begin{matrix} [\alpha] \\ G_n^{*r} \\ [a,r,s_1,\cdots,s_\beta] \\ [@,2,j_1,\cdots,j_\beta] \end{matrix}$$

$$[S]_n \begin{bmatrix} [\alpha] \\ G_n^{*r} \\ [a,r,s_1,\cdots,s_\beta] \\ [@,2,j_1,\cdots,j_\beta] \end{bmatrix} = a \left[\frac{r^{(n+1)}[n(r-1)-1] - [n^\alpha(r-1)^\alpha - 1]}{(r-1)^\alpha[n(r-1)-1]} \right]$$

$$\sum_{\phi=1}^{\beta} \sum_{i=j_\phi}^{(n-1)} \left(\frac{(n-i)(n-i+1)\cdots(n-i+\phi+\alpha-2)}{(\phi+\alpha-1)!} \right) s_\phi^{\left(\frac{(n-i)(n-i+1)}{2!} \right)}$$

$$= \begin{matrix} [\alpha+1] \\ G_n^{*r} \\ [a,r,s_1,\cdots,s_\beta] \\ [@,2,j_1,\cdots,j_\beta] \end{matrix}$$

$$[S]_n \begin{bmatrix} [\alpha+1] \\ G_n^{*r} \\ [a,r,s_1,\cdots,s_\beta] \\ [@,2,j_1,\cdots,j_\beta] \end{bmatrix} = a \left[\frac{r^{(n+1)}[n(r-1)-1] - [n^{(\alpha+1)}(r-1)^{(\alpha+1)} - 1]}{(r-1)^{(\alpha+1)}[n(r-1)-1]} \right]$$

$$\sum_{\phi=1}^{\beta} \sum_{i=j_\phi}^{(n-1)} \left(\frac{(n-i)(n-i+1)\cdots(n-i+\phi+\alpha-1)}{(\phi+\alpha)!} \right) s_\phi^{\left(\frac{(n-i)(n-i+1)}{2!} \right)}$$

$$= \begin{matrix} [\alpha+2] \\ G_n^{*r} \\ [a,r,s_1,\cdots,s_\beta] \\ [@,2,j_1,\cdots,j_\beta] \end{matrix}$$
------- $\ggg\ggg$ $\infty\infty$ FOR EVER ∞

Crazy–Geometric Progressions

The following possibilities may be noted.

$$\left[\begin{array}{c} CG_n^* \\ [a, r_1, r_2, \cdots, r_\beta] \end{array} \right] = a r_1^{(n-1)} r_2^{\left(\frac{(n-1)(n)}{2!} \right)} r_3^{\left(\frac{(n-1)(n)(n+1)}{3!} \right)} \cdots$$

$$r_\beta^{\left(\frac{(n-1)(n)(n+1) \cdots (n+\beta-2)}{\beta!} \right)} \qquad [n \geq 1]$$

"a" as the Initiating Term, "r_1", "r_2", "r_3", $\cdots$, "r_β" as the Inductive Ratios [a, $r_1, r_2, r_3, \cdots, r_\beta$ are algebraic numbers $r_1 \neq 1$].

We may call this type of Progression THE CRAZY–GEOMETRIC PROGRESSSION. Analytic Formulae for the Super–Sums do not allow for any possible simplification in these cases. (See [1])

We could of course induct and repeat the same inductive–block at as many random points of entry and generalize appropriately. The details though trivial are cumbersome. (See [1])

Since the initiating term and Common Ratio and the Inductive Ratios can be any algebraic number, we can see that each family of the Progressions elucidated here, defines a unique Algebraic Field of Sequences.

References

[1] Narayanan Raghunathan:- *Functions and their Progressions – An Elementary Text*. [unpublished]

The Multinomial Theorems

(Dedicated to my friend Shaikh Kamran)

Abstract. The generalizations of the Classical binomial theorem is done here.

Key Words:- Binomial theorem, Multinomial theorems, Generalizations
AMS Subject Classification No:- 11A99

Generalizing the Binomial Theorem we have the following results.

Theorem 0.1 (Basic Multinomial Theorem). *For* $\epsilon \geq 2$

$$
\begin{aligned}
(a_1 + a_2 + a_3 + \cdots + a_\epsilon)^n &= \left[\sum_{k_1=0}^{n} \binom{n}{k_1} a_1^{(n-k_1)} \left[\sum_{k_2=0}^{k_1} \binom{n}{k_2} a_2^{(k_1-k_2)} \right. \right. \\
&\quad \left[\sum_{k_3=0}^{k_2} \binom{n}{k_3} a_3^{(k_2-k_3)} \cdots \right. \\
&\quad \left. \left. \left. \left[\sum_{k_{\epsilon-1}=0}^{k_{\epsilon-2}} \binom{n}{k_{\epsilon-1}} a_{\epsilon-1}^{(k_{\epsilon-2}-k_{\epsilon-1})} a_\epsilon^{(k_{\epsilon-1})} \right] \right] \cdots \right] \\
&= \left[\sum_{k_1=0}^{n} \binom{n}{k_1} a_1^{(n-k_1)} \right. \\
&\quad \left. \left[\prod_{i=2}^{\epsilon-1} \left[\sum_{k_i=0}^{k_{i-1}} \binom{n}{k_i} a_i^{(k_{i-1}-k_i)} a_\epsilon^{(k_{\epsilon-1})} \right] \right] \right]
\end{aligned}
$$

Proof. Appropriate substitutions are Repeated application of the Binomial Theorem yields the Result. The result is valid for All Algebraic values of "a_1". I am told by a friend Shri Ram that the above result is worked out in the Algebra book by Hall and Knight Vol 3. But I have stated it here for completeness in this chapter. But The following results are perhaps not noted so far. □

Theorem 0.2 (Exponential Multinomial Theorem). *For $\epsilon \geq 2$*

$$(a_1 + a_2)^{n_1} + a_3)^{n_2} + \cdots + a_\epsilon)^{n_{\epsilon-1}} = \left[\sum_{k_{\epsilon-1}=0}^{n_{\epsilon-1}} \binom{n_{\epsilon-1}}{k_{\epsilon-1}} \left[\left[\cdots \cdots \right. \right. \right.$$

$$\left[\sum_{k_3=0}^{n_3} \binom{n_3}{k_3} \left[\sum_{k_2=0}^{n_2} \binom{n_2}{k_2} \left[\sum_{k_1=0}^{n_1} \binom{n_1}{k_1} a_1^{(n_1-k_1)} a_2^{k_1} \right]^{(n_2-k_2)} a_3^{k_2} \right]^{(n_3-k_3)} a_4^{k_3}$$

$$\left. \left. \cdots \cdots \right] \right]^{(n_{\epsilon-1}-k_{\epsilon-1})} a_\epsilon^{k_{\epsilon-1}} \right]$$

Proof. Appropriate substitutions and Repeated application of the Binomial Theorem yields the Result.

The result is valid for All Algebraic values of "a_i". □

Theorem 0.3 (Generalized Exponential Multinomial Theorem). *Let*

$$[A_t^\gamma] = \left[\sum_{{}^t k_1=0}^{\gamma} \binom{\gamma}{{}^t k_1} {}^t a_1^{(\gamma - {}^t k_1)} \left[\prod_{i=2}^{\epsilon_t-1} \left[\sum_{{}^t k_i=0}^{{}^t k_{i-1}} \binom{\gamma}{{}^t k_i} {}^t a_i^{({}^t k_{i-1} - {}^t k_i)} {}^t a_{\epsilon_t}^{({}^t k_{\epsilon_t} - 1)} \right] \right] \right]$$

Then

For $\epsilon_i \geq 1$ for all $(1 \leq i \leq \tau)$ and $\tau \geq 1$

$$\left({}^1 a_1 + {}^1 a_2 + \cdots + {}^1 a_{\epsilon_1} \right)^{n_1} + {}^2 a_1 + {}^2 a_2 + \cdots + {}^2 a_{\epsilon_2})^{n_2} +)) \cdots)$$

$$+ {}^\tau a_1 + {}^\tau a_2 + \cdots + {}^\tau a_{\epsilon_\tau})^{n_\tau}$$

$$= \left[\sum_{k_\tau=0}^{n_\tau} \binom{n_\tau}{k_\tau} \left[\left[\cdots \cdots \right. \right. \right.$$

$$\left[\sum_{k_3=0}^{n_3} \binom{n_3}{k_3} \left[\sum_{k_2=0}^{n_2} \binom{n_2}{k_2} [A_1^{n_1}]^{(n_2-k_2)} [A_2^{n_2}]^{k_2} \right]^{(n_3-k_3)} [A_3^{n_3}]^{k_3} \right]^{(n_4-k_4)} [A_4^{n_4}]^{k_4}$$

$$\left. \left. \cdots \cdots \right] \right]^{(n_\tau-k_\tau)} [A_\tau^{n_\tau}]^{k_\tau} \right]$$

Proof. Appropriate substitutions and Repeated application of the Binomial Theorem yields the Result.

The result is valid for All Algebraic values of "${}^t a_i$". □

References

[1] David M. Burton, *ELEMENTARY NUMBER THEORY*, UNIVERSAL BOOK STALL. NEW DELHI. Second Edition. Reprint 1998.

Triangular Numbers – Some General Theorems and Related Results

(Dedicated to my esteemed teacher the inimitable N. B. N. – SSKZM)

Abstract. The following results are proved.

If $\mathbf{T} = \{T_0, T_1, T_2, T_3, \cdots T_n, \cdots, \infty\}$ $\quad [T_0 = 0]$ are Triangular numbers

1. then $(2\xi + 1)^2 T_n + T_\xi$ $\quad [\xi \geq 0, n \geq 0]$ are also Triangular Numbers.
2. then $(2\lambda)^2 T_n + T_\lambda + \lambda n [\lambda \geq 1, n \geq 0]$ are also Triangular numbers.
3. The square of every odd number can be expressed as the difference of two Triangular Numbers.
4. The square of every λ^{th} even number can be expressed as the difference of the '$(3\lambda)^{\text{th}}$ Triangular Number' and 'the sum of the λth Triangular Number and λ'.
5. For any Natural Number $N \geq 2$, there Exist Infinite Triangular Numbers that are the Sum of N Triangular Numbers each.
6. If
$$\mathbf{x} = \frac{n(n + 2\alpha + 1)}{2} + 1, \; y = n + 1, \; z = \frac{n(n + 2\alpha + 1)}{2}$$
$$\mathbf{T_x} = T_y + T_z + (\alpha - 1)n = T_y + T_z + (\alpha - 1)(y - 1)$$
$$[\alpha = 0, 1, 2, \cdots \infty] \, [n = 0, 1, 2, \cdots, \infty]$$

7. If
$$\mathbf{x} = \frac{(n + 1)(n + 2\beta)}{2} + 1, \; y = n + 1, \; z = \frac{(n + 1)(n + 2\beta)}{2}$$
$$\mathbf{T_x} = T_y + T_z + \beta + (\beta - 1)n = T_y + T_z + \beta$$
$$+ (\beta - 1)(y - 1)[\beta = 1, 2, 3 \cdots \infty] \, [n = 0, 1, 2, \cdots, \infty]$$

Various related sub-results are also proved.

Infinite Types of unique sub-classes of Triangular Numbers are defined and the formulae or their sum upto "n" terms are derived and these are expressed in terms of binomial coefficients also.

Key Words:- Triangular Numbers, Binomial Coefficiets, Infinite Generalizations.

AMS Subject Classification No:- 11A99.

1. Some General Results

Theorem 1.1. *If* $\mathbf{T} = \{T_0, T_1, T_2, T_3, \cdots T_n, \cdots \infty\}[T_0 = 0]$ *are Triangular Numbers then* $(2\xi + 1)^2 T_n + T_\xi$ $[\xi \geq 0, n \geq 0]$ *are also Triangular Numbers.*

[Euler's results that $(9T_n + 1), (25T_n + 3), (49T_n + 6), [n \geq 0]$ *are Triangular, are obtained as special cases when we set* $\xi = 1, 2,$ *and 3 respectively in the formula* $(2\xi + 1)^2 T_n + T_\xi.]$

Proof. We prove the result by repeated induction.

For $\xi = 0, (2\xi + 1)^2 T_n + T_\xi$ are the set of Triangular Numbers themselves. For $\xi = 1, 2, 3$ we have the standard results due to Euler which are proved by induction on "n" for each specific value of ξ. To prove the general formula for all values of $\xi \geq 0$ we assume that the result is true for $\xi = k$ and prove that it is true for $\xi = k + 1$ and thus close the doubly (infinitely) inductive argument. We state the complete set of Triangular Numbers for each value of ξ as follows which can easily be proved by induction and the Pythagorean. Theorem that a number is Triangular if and only if it is of the form $\dfrac{n(n+1)}{2}$, in each case.

[Note on the notation:

$^{[R]}$

In $^L T_n$, R may be called the Root Level and L the Layer level which define a specific sub-class of Triangular Numbers.]

For $\xi = 0, 2\xi + 1 = 1$

$$^1_1\mathbf{T_n}^{[1]} = \frac{n(n+1)}{2} = T_n + T_0 = (T_0 + T_1)T_n + T_0 = T_n \quad [n \geq 1]$$

For $\xi = 1, 2\xi + 1 = 3$

$$^1_1\mathbf{T_n}^{[3]} = \frac{(3n+1)(3n+2)}{2} = 9T_n + T_1 = (T_2 + T_3)T_n + T_1$$
$$= T_{(3n+1)} \quad [n \geq 1]$$

For $\xi = 2, 2\xi + 1 = 5$

$$^3_2\mathbf{T_2}^{[3]} = \frac{(6n+1)(6n+2)}{2} = 25T_n + T_2 = (T_4 + T_5)T_n + T_2$$
$$= T_{(6n+1)} \quad [n \geq 1]$$

For $\xi = 3, 2\xi + 1 = 7$

$$^3_3\mathbf{T_3}^{[3]} = \frac{(9n+1)(9n+2)}{2} = 49T_n + T_3 = (T_6 + T_7)T_n + T_3$$
$$= T_{(9n+1)} \quad [n \geq 1]$$

For $\xi = 4, 2\xi + 1 = 9$

$$^3_4\mathbf{T_n}^{[3]} = \frac{(12n+1)(12n+2)}{2} = 81T_n + T_4$$
$$= (T_8 + T_9)T_n + T_4 = T_{(12n+1)} \quad [n \geq 1]$$

For $\xi = 5, 2\xi + 1 = 11$

$$^5\mathbf{T_n^{[3]}} = \frac{(15n+1)(15n+2)}{2} = 121T_n + T_5$$
$$= (T_{10} + T_{11})T_n + T_5 = T_{(15n+1)} \quad [n \geq 1]$$

For $\xi = 6, 2\xi + 1 = 13$

$$^6\mathbf{T_n^{[3]}} = \frac{(18n+1)(18n+2)}{2} = 169T_n + T_6$$
$$= (T_{12} + T_{13})T_n + T_6 = T_{(18n+1)} \quad [n \geq 1]$$

Clearly if $\xi = k$

$$^k\mathbf{T_n^{[3]}} = \frac{(3kn+1)(3kn+2)}{2} = (2k+1)^2 T_n + T_k$$
$$= (T_{2k} + T_{2k+1}) + T_n + T_k = T_{(3kn+1)} \quad [n \geq 1])$$

are Triangular, then for $\xi = k + 1$

$$\mathbf{k+1} \mathbf{T_n^{[3]}} = \frac{(3(k+1)n+1)(3(k+1)n+2)}{2}$$
$$= (2(k+1)+1)^2 T_n + T_{k+1}$$
$$= (T_{2(k+1)} + T_{2(k+1)+1}) + T_n + T_{k+1}$$
$$= T_{(3(k+1)n+1)} \quad [n \geq 1]$$

are also Triangular thus proving that

$$\xi \mathbf{T_n^{[3]}} = \frac{(3\xi n+1)(3\xi n+2)}{2} = (2\xi+1)^2 T_n + T_\xi$$
$$= (T_{2\xi} + T_{2\xi+1})T_n + T_\xi = T_{(3\xi n+1)} \quad [\xi \geq 1, n \geq 1] \quad (1.1)$$

are all Triangular. Thus we may say that Triangular Numbers derived form Root Level 3 Layer Levels $\xi[\xi \geq 1]$ are Eulerian classes of Triangular Numbers and we may call "ξ" the Eulerian Class Level Number. For $\xi = 0$ the Eulerian class is the set of all Triangular Numbers. $\qquad\square$

Corollary 1.2. Every Triangular Number can be Expressed the Difference of a Triangular Number and the Sum of two Triangular numbers and the product of another Triangular Number in Infinite Ways.

$$\mathbf{T_\xi} = T_{(3\xi n+1)} - (T_{2\xi} + T_{2\xi+1})T_n \qquad [\xi \geq 1, n \geq 1]$$

Proof. Trivial from (1.1). $\qquad\square$

Corollary 1.3. Every Triangular Number can be Expressed as the Ratio of the Difference of Two Triangular Numbers and the Sum of Two Triangular Numbers in Infinite Ways.

$$\mathbf{T_n} = \frac{T_{(3\xi n+1)} - T_\xi}{(T_{2\xi} + T_{2\xi+1})} \qquad [\xi \geq 1, n \geq 1]$$

Proof. Trivial form (1.1).   □

Corollary 1.4. For $\xi \geq 1, n \geq 1$

$$(\mathbf{2\xi + 1})^2(\mathbf{n + 1})^2 = T_{(3\xi n+1)} + T_{(3\xi(n+1)+1)} - 2T_\xi$$

Proof.

$$\overset{[\mathbf{3}]}{\xi}\mathbf{T_n} = \frac{(3\xi n + 1)(3\xi n + 2)}{2} = (2\xi + 1)^2 T_n + T_\xi$$
$$= (T_{2\xi} + T_{2\xi+1})T_n + T_\xi = T_{(3\xi n+1)} \quad [\xi \geq 1, n \geq 1]$$

$$(1.2)$$

$$(\mathbf{2\xi + 1})^2\mathbf{T_n} + \mathbf{T_\xi} = T_{(3\xi n+1)}. \tag{1.3}$$

Substituting $n = n + 1$ in (1.3)

$$(\mathbf{2\xi + 1})^2\mathbf{T_{n+1}} + \mathbf{T_\xi} = T_{(3\xi(n+1)+1)} \tag{1.4}$$

adding the two equations, substituting $T_n + T_{n+1} = (n+1)^2$ [Nicomachus] and simplifying we have

$$(\mathbf{2\xi + 1})^2(\mathbf{n + 1})^2 = T_{(3\xi n+1)} + T_{(3\xi(n+1)+1)} - 2T_\xi \quad [\xi \geq 1, n \geq 1] \tag{1.5}$$

$2\xi + 1$ defines the set of all odd numbers. Seeking a symmetric result for all even numbers, we have the following theorem.   □

Theorem 1.5. *If* $T = \{T_0, T_1, T_2, T_3, \cdots T_n, \cdots \infty\}[T_0 = 0]$ *are Triangular Numbers then* $(2\lambda)^2 T_n + T_\lambda + \lambda n$ $[\lambda \geq 1, n \geq 0]$ *are also Triangular Numbers.*

Proof. We prove the result by repeated induction.
First we prove for $\lambda = 1, 2, 3$ by induction on "n" for each specific value of λ. To prove the general formula for all values of $\lambda \geq 1$ we assume that the result is true for $\lambda = k$ and prove that it is true for $\lambda = k + 1$ and thus close the doubly (infinitely) inductive argument. We state the complete set of Triangular Numbers for each value of λ as follows which can easily be proved by induction and the Pythagorean Theorem that a number is Triangular if and only if if it is the form $\frac{n(n+1)}{2}$, in each case.
For $\lambda = 1, \ 2\lambda = 2$

$$\overset{[\mathbf{1}]}{\mathbf{2}}\mathbf{T_n} = \frac{(2n+1)(2n+2)}{2} = 4T_n + T_1 + 1n$$
$$= (T_1 + T_2)T_n + T_1 + 1(n) = T_{3n} \quad [n \geq 1]$$

For $\lambda = 2,\ 2\lambda = 4$

$$5^{[1]}_{\mathbf{T_n}} = \frac{(5n+1)(5n+2)}{2} = 16T_n + T_2 + 2n$$
$$= (T_3 + T_4)T_n + T_2 + 2n = T_{6n} \quad [n \geq 1]$$

For $\lambda = 3,\ 2\lambda = 6$

$$8^{[1]}_{\mathbf{T_n}} = \frac{(8n+1)(8n+2)}{2} = 36T_n + T_3 + 3n$$
$$= (T_5 + T_6)T_n + T_3 + 3n = T_{9n} \quad [n \geq 1]$$

For $\lambda = 4,\ 2\lambda = 8$

$$11^{[1]}_{\mathbf{T_n}} = \frac{(11n+1)(11n+2)}{2} = 64T_n + T_4 + 4n$$
$$= (T_7 + T_8)T_n + T_4 + 4n = T_{12n} \quad [n \geq 1]$$

For $\lambda = 5,\ 2\lambda = 10$

$$14^{[1]}_{\mathbf{T_n}} = \frac{(14n+1)(14n+2)}{2} = 100T_n + T_5 + 5n$$
$$= (T_9 + T_{10})T_n + T_5 + 5n = T_{15n}$$

For $\lambda = 6,\ 2\lambda = 12$

$$17^{[1]}_{\mathbf{T_n}} = \frac{(17n+1)(17n+2)}{2} = 144T_n + T_6 + 6n$$
$$= (T_{11} + T_{12})T_n + T_6 + 6n = T_{18n} \quad [n \geq 1]$$

For $\lambda = 7,\ 2\lambda = 14$

$$20^{[1]}_{\mathbf{T_n}} = \frac{(20n+1)(20n+2)}{2} = 196T_n + T_7 + 7n$$
$$= (T_{13} + T_{14})T_n + T_7 + 7n = T_{21n} \quad [n \geq 1]$$

For $\lambda = k,\ 2\lambda = 2k$

$$(\mathbf{3k-1})^{[1]}_{\mathbf{T_n}} = \frac{((3k-1)n+1)((3k-1)n+2)}{2}$$
$$= (2k)^2 T_n + T_k + kn$$
$$= (T_{k-1} + T_k)T_n + T_k + kn = T_{3kn} \quad [n \geq 1]$$

For $\lambda = k+1, \ 2\lambda = 2(k+1)$

$$
\begin{aligned}
(3\mathbf{k}+2)\mathbf{T_n}^{[1]} &= \frac{((3k+2)n+1)((3k+2)n+2)}{2} \\
&= (2(k+1))^2 T_n + T_{(k+1)} + (k+1)n \\
&= (T_k + T_{k+1})T_n + T_{(k+1)} + (k+1)n = T_{(3k+1)n} \quad [n \geq 1]
\end{aligned}
$$

are also triangular thus proving that

$$
\begin{aligned}
(3\lambda-1)\mathbf{T_n}^{[1]} &= \frac{((3\lambda-1)n+1)((3\lambda-1)n+2)}{2} \\
&= (2\lambda)^2 T_n + T_\lambda + \lambda n \\
&= (T_{2\lambda-1} + T_{2\lambda})T_n + T_\lambda + \lambda n = T_{3\lambda n} \quad [n \geq 1]
\end{aligned}
\tag{1.6}
$$

are all Triangular. Thus we may say that Triangular Numbers derived from Root Level 1 Layer Levels $(3\lambda - 1)[\lambda \geq 1]$ are Pythagorean classes of Triangular Numbers and we may call "$(3\lambda - 1)$" the Pythagorean Class Level Number. □

Corollary 1.6. Every Triangular Number can be Expressed the "Difference of a Triangular Number" and "the Sum of Two Triangular Numbers and the product of another Triangular Number and a specific constant (product of the two indexing factors)" in Infinite Ways.

$$\mathbf{T_\lambda} = T_{3\lambda n} - (T_{2\lambda-1} + T_{2\lambda})T_n - \lambda n \qquad [\lambda \geq 1, n \geq 1]$$

Proof. Trivial from (1.6) □

Corollary 1.7. Every Triangular Number can be Expressed as the Ratio of the "Difference of a Triangular Number and the Sum of a Triangular Number and a specific constant (product of the two indexing factors)" and the "Sum of Two Triangular Numbers" in Infinite Ways.

$$\mathbf{T_\lambda} = T_{3\lambda n} - (T_{2\lambda-1} + T_{2\lambda})T_n - \lambda n$$

$$\mathbf{T_n} = \frac{T_{3\lambda n} - (T_\lambda + \lambda n)}{(T_{2\lambda-1} + T_{2\lambda})} \qquad [\lambda \geq 1, n \geq 1]$$

Proof. Trivial from (1.6) □

Corollary 1.8. For $\lambda \geq 1, n \geq 1$

$$(2\lambda)^2(\mathbf{n}+1)^2 = T_{3\lambda n} + T_{(3\lambda(n+1))} - (2T_\lambda + \lambda(2n+1)). \tag{1.7}$$

Proof.

$$(3\lambda - 1)^{[1]} \mathbf{T_n} = \frac{((3\lambda - 1)n + 1)((3\lambda - 1)n + 2)}{2}$$
$$= (2\lambda)^2 T_n + T_\lambda + \lambda n \qquad (1.8)$$
$$= (T_{2\lambda - 1} + T_{2\lambda})T_n + T_\lambda + \lambda n = T_{3\lambda n} \quad [\lambda \geq 1, n \geq 1]$$

$$(2\lambda)^2 \mathbf{T_n} + \mathbf{T_\lambda} + \lambda \mathbf{n} = T_{3\lambda n} \qquad (1.9)$$

Substituting $n = n + 1$ in (1.9)

$$(2\lambda)^2 \mathbf{T_{n+1}} + \mathbf{T_\lambda} + \lambda(\mathbf{n} + 1) = T_{(3\lambda(n+1))} \qquad (1.10)$$

adding the two equations substituting $T_n + T_{n+1} = (n+1)^2$ [Nicomachus] and simplifying we have

$$(2\lambda)^2 (\mathbf{n} + 1)^2 = T_{3\lambda n} + T_{(3\lambda(n+1))} - (2T_\lambda + \lambda(2n+1))[\lambda \geq 1, n \geq 1] \quad (1.11)$$

$\square$

2. Two General Theorems on Triangular Numbers

Theorem 2.1. *The square of every odd number can be expressed as the difference of two Triangular Numbers.*

 Proof. We have

$$(2\xi + 1)^2 \mathbf{T_n} + \mathbf{T_\xi} = T_{(3\xi n + 1)} \qquad (2.1)$$

Substituting $n = 1$ in (2.1), we have

$$(2\xi + 1)^2 = T_{(3\xi + 1)} - T_\xi \quad [\xi \geq 0] \qquad (2.2)$$

The standard result that "the square of any odd multiple of 3 is the difference of two Triangular Numbers [i.e., $9(2n+1)^2 = T_{9n+4} - T_{3n+1}$]" [Ref. 6], clearly is a cumbersome expression of a special case when the "odd multiples of 3" are the odd numbers concerned! Again, looking for General Symmetry we have. $\square$

Theorem 2.2. *The square of every λ^{th} even number can be expressed as the difference of the '$(3\lambda)^{th}$, Triangular Number' and 'the sum of the λth Triangular Number and λ'.*

 Proof. We have,

$$(2\lambda)^2 \mathbf{T_n} + \mathbf{T_\lambda} + \lambda \mathbf{n} = T_{3\lambda n} \qquad (2.3)$$

Substituting $n = 1$ in (2.3), we have

$$(2\lambda)^2 = T_{3\lambda} - (T_\lambda + \lambda) \quad [\lambda \geq 1] \qquad (2.4)$$

$\square$

3. Infinite Root Levels and Layer Levels of Triangular Numbers

In this section we define various Root-Levels and Layer-Levels of Triangular Numbers and investigate them.

In ${}^{L}_{R}T_n$, R may be called the Root Level and L the Layer Level which define a specific sub-class of Triangular Numbers.

The sequence of Layer-Levels for each Root-Level Facilitates the Proof for All Root-Levels.

$${}^{[1]}_{1}T_{n+1} = \frac{(n+1)(n+2)}{2} \qquad\qquad [n = 0, 1, 2, \cdots, \infty]$$
$$= T_1, T_2, T_3, T_4, T_5, \cdots, T_{1+(n-1)}, \cdots, \infty$$

$${}^{[1]}_{2}T_{n+1} = \frac{(2n+1)(2n+2)}{2} \qquad\qquad [n = 0, 1, 2, \cdots, \infty]$$
$$= T_1, T_3, T_5, T_7, T_9, \cdots, T_{1+2(n-1)}, \cdots, \infty$$

$${}^{[1]}_{3}T_{n+1} = \frac{(3n+1)(3n+2)}{2} \qquad\qquad [n = 0, 1, 2, \cdots, \infty]$$
$$= T_1, T_4, T_7, T_{10}, T_{13}, \cdots, T_{1+3(n-1)}, \cdots, \infty$$

$${}^{[1]}_{4}T_{n+1} = \frac{(4n+1)(4n+2)}{2} \qquad\qquad [n = 0, 1, 2, \cdots, \infty]$$
$$= T_1, T_5, T_9, T_{13}, T_{17}, \cdots, T_{1+4(n-1)}, \cdots, \infty$$

— — — — — — — — —

$${}^{[1]}_{\phi}T_{n+1} = \frac{(\phi n+1)(\phi n+2)}{2} \qquad\qquad [n = 0, 1, 2, \cdots, \infty]$$
$$= T_1, T_{1+\phi}, T_{1+2\phi}, T_{1+3\phi}, T_{1+4\phi}, \cdots, T_{1+\phi(n-1)}, \cdots, \infty$$

$${}^{[2]}_{1}T_{n+1} = \frac{(2n+1)(2n+2)}{2} \qquad\qquad [n = 0, 1, 2, \cdots, \infty]$$
$$= T_1, T_3, T_5, T_7, T_9, \cdots, T_{1+2(n-1)}, \cdots, \infty$$

$${}^{[2]}_{2}T_{n+1} = \frac{(4n+1)(4n+2)}{2} \qquad\qquad [n = 0, 1, 2, \cdots, \infty]$$
$$= T_1, T_5, T_9, T_{13}, T_{17}, \cdots, T_{1+4(n-1)}, \cdots, \infty$$

$$3\mathbf{T_{n+1}^{[2]}} = \frac{(6n+1)(6n+2)}{2} \qquad [n=0,1,2,\cdots,\infty]$$

$$= T_1, T_7, T_{13}, T_{19}, T_{25}, \cdots, T_{1+6(n-1)}, \cdots, \infty$$

$$4\mathbf{T_{n+1}^{[2]}} = \frac{(8n+1)(8n+2)}{2} \qquad [n=0,1,2,\cdots,\infty]$$

$$= T_1, T_9, T_{17}, T_{25}, T_{33}, \cdots, T_{1+8(n-1)}, \cdots, \infty$$

$-\ -\ -\ -\ -\ -\ -\ -\ -\ -\ -$

$$\phi\mathbf{T_{n+1}^{[2]}} = \frac{(2\phi n+1)(2\phi n+2)}{2} \qquad [n=0,1,2,\cdots,\infty]$$

$$= T_1, T_{1+2\phi}, T_{1+4\phi}, T_{1+6\phi}, T_{1+8\phi}, \cdots, T_{1+(2\phi n-1)}, \cdots, \infty$$

$$1\mathbf{T_{n+1}^{[3]}} = \frac{(3n+1)(3n+2)}{2} \qquad [n=0,1,2,\cdots,\infty]$$

$$= T_1, T_4, T_7, T_{10}, T_{13}, \cdots, T_{1+3(n-1)}, \cdots, \infty$$

$$2\mathbf{T_{n+1}^{[3]}} = \frac{(6n+1)(6n+2)}{2} \qquad [n=0,1,2,\cdots,\infty]$$

$$= T_1, T_7, T_{13}, T_{19}, T_{25}, \cdots, T_{1+6(n-1)}, \cdots, \infty$$

$$3\mathbf{T_{n+1}^{[3]}} = \frac{(9n+1)(9n+2)}{2} \qquad [n=0,1,2,\cdots,\infty]$$

$$= T_1, T_{10}, T_{19}, T_{28}, T_{37}, \cdots, T_{1+9(n-1)}, \cdots, \infty$$

$$4\mathbf{T_{n+1}^{[3]}} = \frac{(12n+1)(12n+2)}{2} \qquad [n=0,1,2,\cdots,\infty]$$

$$= T_1, T_{13}, T_{25}, T_{37}, T_{49}, \cdots, T_{1+12(n-1)}, \cdots, \infty$$

$-\ -\ -\ -\ -\ -\ -\ -$

$$\phi\mathbf{T_{n+1}^{[3]}} = \frac{(3\phi n+1)(3\phi n+2)}{2} \qquad [n=0,1,2,\cdots,\infty]$$

$$= T_1, T_{1+3\phi}, T_{1+6\phi}, T_{1+9\phi}, T_{1+12\phi}, \cdots, T_{1+3\phi(n-1)}, \cdots, \infty$$

$$1\mathbf{T_{n+1}^{[4]}} = \frac{(4n+1)(4n+2)}{2} \qquad [n=0,1,2,\cdots,\infty]$$

$$= T_1, T_5, T_9, T_{13}, T_{17}, \cdots, T_{1+4(n-1)}, \cdots, \infty$$

$$2\mathbf{T_{n+1}^{[4]}} = \frac{(8n+1)(8n+2)}{2} \qquad [n=0,1,2,\cdots,\infty]$$

$$= T_1, T_9, T_{17}, T_{25}, T_{33}, \cdots, T_{1+8(n-1)}, \cdots, \infty$$

$${}_{3}^{[4]}\mathbf{T_{n+1}} = \frac{(12n+1)(12n+2)}{2} \qquad\qquad [n = 0, 1, 2, \cdots, \infty]$$

$$= T_1, T_{13}, T_{25}, T_{37}, T_{49}, \cdots, T_{1+12(n-1)}, \cdots, \infty$$

$${}_{4}^{[4]}\mathbf{T_{n+1}} = \frac{(16n+1)(16n+2)}{2} \qquad\qquad [n = 0, 1, 2, \cdots, \infty]$$

$$= T_1, T_{17}, T_{33}, T_{49}, T_{65}, \cdots, T_{1+16(n-1)}, \cdots, \infty$$

$- - - - - - - - -$

$${}_{\phi}^{[4]}\mathbf{T_{n+1}} = \frac{(\phi 4n+1)(\phi 4n+2)}{2} \qquad\qquad [n = 0, 1, 2, \cdots, \infty]$$

$$= T_1, T_{1+4\phi}, T_{1+8\phi}, T_{1+12\phi}, T_{1+16\phi}, \cdots, T_{1+4\phi(n-1)}, \cdots, \infty$$

$- - - - - - - -$

$${}_{1}^{[\mathbf{R}]}\mathbf{T_{n+1}} = \frac{(Rn+1)(Rn+2)}{2} \qquad\qquad [n = 0, 1, 2, \cdots, \infty]$$

$$= T_1, T_{1+R}, T_{1+2R}, T_{1+3R}, T_{1+4R}, \cdots, T_{1+R(n-1)}, \cdots, \infty$$

$${}_{2}^{[\mathbf{R}]}\mathbf{T_{n+1}} = \frac{(2Rn+1)(2Rn+2)}{2} \qquad\qquad [n = 0, 1, 2, \cdots, \infty]$$

$$= T_1, T_{1+2R}, T_{1+4R}, T_{1+6R}, T_{1+8R}, \cdots, T_{1+2R(n-1)}, \cdots, \infty$$

$${}_{3}^{[\mathbf{R}]}\mathbf{T_{n+1}} = \frac{(3Rn+1)(3Rn+2)}{2} \qquad\qquad [n = 0, 1, 2, \cdots, \infty]$$

$$= T_1, T_{1+3R}, T_{1+6R}, T_{1+9R}, T_{1+12R}, \cdots, T_{1+3R(n-1)}, \cdots, \infty$$

$${}_{4}^{[\mathbf{R}]}\mathbf{T_{n+1}} = \frac{(4Rn+1)(4Rn+2)}{2} \qquad\qquad [n = 0, 1, 2, \cdots, \infty]$$

$$= T_1, T_{1+4R}, T_{1+8R}, T_{1+12R}, T_{1+16R}, \cdots, T_{1+4R(n-1)}, \cdots, \infty$$

$- - - - - - - -$

$${}_{\phi}^{[\mathbf{R}]}\mathbf{T_{n+1}} = \frac{(\phi Rn+1)(\phi Rn+2)}{2} \qquad\qquad [n = 0, 1, 2, \cdots, \infty]$$

$$= T_1, T_{1+\phi R}, T_{1+2\phi R}, T_{1+3\phi R}, T_{1+4\phi R}, \cdots, T_{1+\phi R(n-1)}, \cdots, \infty$$

$$[\phi = 1, 2, \cdots \infty]$$

$$[R = 1, 2, \cdots, \infty]$$

4. The Infinite Triangular Number Classes in Terms of Binomial Coefficients

We can easily prove by Repeated Induction the following results.

$$[n = 0, 1, 2, \cdots, \infty]$$

$$_1^{[1]}\mathbf{T_{n+1}} = \binom{n+2}{2} \quad = T_n + 1 \quad \text{[This is the standard result. Ref.6)]}$$

$$_2^{[1]}\mathbf{T_{n+1}} = \binom{2n+2}{2}$$

$$_3^{[1]}\mathbf{T_{n+1}} = \binom{3n+2}{2}$$

$$_4^{[1]}\mathbf{T_{n+1}} = \binom{4n+2}{2}$$

$$- - - - - - -$$

$$_\phi^{[1]}\mathbf{T_{n+1}} = \binom{\phi n+2}{2}$$

$$_1^{[2]}\mathbf{T_{n+1}} = \binom{2n+2}{2}$$

$$_2^{[2]}\mathbf{T_{n+1}} = \binom{4n+2}{2}$$

$$_3^{[2]}\mathbf{T_{n+1}} = \binom{6n+2}{2}$$

$$_4^{[2]}\mathbf{T_{n+1}} = \binom{8n+2}{2}$$

$$- - - - - - -$$

$$_\phi^{[2]}\mathbf{T_{n+1}} = \binom{2\phi n+2}{2}$$

$$_1^{[3]}\mathbf{T_{n+1}} = \binom{3n+2}{2}$$

$${}^{[3]}_{2}\mathbf{T_{n+1}} = \binom{6n+2}{2}$$

$${}^{[3]}_{3}\mathbf{T_{n+1}} = \binom{9n+2}{2}$$

$${}^{[3]}_{4}\mathbf{T_{n+1}} = \binom{12n+2}{2}$$

$$- - - - - - - -$$

$${}^{[3]}_{\phi}\mathbf{T_{n+1}} = \binom{3\phi n+2}{2}$$

$${}^{[4]}_{1}\mathbf{T_{n+1}} = \binom{4n+2}{2}$$

$${}^{[4]}_{2}\mathbf{T_{n+1}} = \binom{8n+2}{2}$$

$${}^{[4]}_{3}\mathbf{T_{n+1}} = \binom{12n+2}{2}$$

$${}^{[4]}_{4}\mathbf{T_{n+1}} = \binom{16n+2}{2}$$

$$- - - - - - - -$$

$${}^{[4]}_{\phi}\mathbf{T_{n+1}} = \binom{4\phi n+2}{2}$$

$$- - - - - - - -$$

The result if True for k ie.

$$\begin{aligned}{}^{[\mathbf{k}]}_{\phi}\mathbf{T_{n+1}} &= \binom{k\phi n+2}{2} \Rightarrow {}^{[k+1]}_{\phi}T_{n}+1 \\ &= \binom{(k+1)\phi n+2}{2}\end{aligned}$$

is also True for $k+1$ proving the General Result for $R = 1, 2, \cdots \infty$

$${}^{[\mathbf{R}]}_{1}\mathbf{T_{n+1}} = \binom{Rn+2}{2}$$

$$\overset{[\mathbf{R}]}{_2\mathbf{T_{n+1}}} = \binom{2Rn+2}{2}$$

$$\overset{[\mathbf{R}]}{_3\mathbf{T_{n+1}}} = \binom{3Rn+2}{2}$$

$$\overset{[\mathbf{R}]}{_4\mathbf{T_{n+1}}} = \binom{4Rn+2}{2}$$

$$----- --$$

$$\overset{[\mathbf{R}]}{_\phi\mathbf{T_{n+1}}} = \binom{\phi Rn+2}{2}$$

$$[\phi = 1, 2, \cdots, \infty]$$

$$[R = 1, 2, \cdots, \infty]$$

5. The Formulae for the Sum up to the nth Term of Infinite Root Levels and Layer Levels of Triangular Numbers

In this section we derive the Formulae for the Sum up to the nth term of Infinite Root Levels and Layer Levels of Triangular Numbers. Aryabhatta's traditional formula for the Sum of Triangular Numbers is assumed.

For this purpose we need to consider a simple case of General Arithmetic cum Arithmetic Inductive Progressions.   [Ref.7]

Let

$$[\mathbf{P}]_\mathbf{n} = a + b(n-1) + \frac{c(n-1)(n-2)}{2} \tag{5.1}$$

[a, b and c are algebraic numbers.]

[Substituting $a = 1, b = 2$ and $c = 1$ we get the sequence of the set of All Triangular Numbers **T**.]

Now we can easily prove by Induction that the sum of this progression up to n terms is

$$\mathbf{S_n[P]} = an + b\left[\frac{n(n-1)}{2}\right]$$
$$+ \frac{c(n-1)(n-2)}{2} + \frac{2c(n-2)(n-3)}{2}$$
$$+ \frac{3c(n-1)(n-2)}{2} \cdots \frac{(n-2)c(2)(1)}{2}.$$

$$= an + b\left[\frac{n(n-1)}{2}\right] + c\sum_{1}^{n-2} i\frac{(n-i)(n-(i+1))}{2}$$

$$= an + b\left[\frac{n(n-1)}{2!}\right] + c\left[\frac{n(n-1)(n-2)}{3!}\right]$$

$$= \binom{a}{1}\binom{n}{1} + \binom{b}{1}\binom{n}{2} + \binom{c}{1}\binom{n}{3} \qquad (5.2)$$
$$\underbrace{\boxed{n \geq 1}} \qquad \underbrace{\boxed{n \geq 2}} \qquad \underbrace{\boxed{n \geq 3}}$$

$$\sum_{0}^{n-1} {}^{1}T_{i+1}^{[1]} = \sum_{0}^{n-1} \frac{(i+1)(i+2)}{2} \qquad [n = 1, 2, \cdots, \infty]$$

$$= \sum_{1}^{n} T_{1+(i-1)} = n + (T_2 - T_1)\left[\frac{n(n-1)}{2}\right]$$

$$+ 1^2 \sum_{1}^{n-2} i\frac{(n-i)(n-(i+1))}{2}$$

Substituting $a = 1, b = 2, c = 1^2 = 1$, in (5.2) we have

$$= n + 2\left[\frac{n(n-1)}{2}\right] + \sum_{1}^{n-2} i\frac{(n-i)(n-(i+1))}{2}$$

$$= n + 2\left[\frac{n(n-1)}{2!}\right] + \left[\frac{n(n-1)(n-2)}{3!}\right]$$

$$= \binom{n}{1} + \binom{2}{1}\binom{n}{2} + \binom{n}{3}$$
$$\underbrace{\boxed{n \geq 1}} \qquad \underbrace{\boxed{n \geq 2}} \qquad \underbrace{\boxed{n \geq 3}}$$

$$\sum_{0}^{n-1} {}^{2}T_{i+1}^{[1]} = \sum_{0}^{n-1} \frac{(2i+1)(2i+2)}{2} \qquad [n = 1, 2, \cdots, \infty]$$

$$= \sum_{1}^{n} T_{1+2(i-1)}$$

$$= n + (T_3 - T_1)\left[\frac{n(n-1)}{2}\right] + 2^2 \sum_{1}^{n-2} i\frac{(n-i)(n-(i+1))}{2}.$$

Substituting $a = 1, b = 5, c = 2^2 = 4$, in (5.2) we have

$$= n + 5\left[\frac{n(n-1)}{2}\right] + 4\sum_{1}^{n-2} i\frac{(n-i)(n-(i+1))}{2}$$

$$= n + 5 \left[\frac{n(n-1)}{2!} \right] + 4 \left[\frac{n(n-1)(n-2)}{3!} \right]$$

$$= \underbrace{\binom{n}{1}}_{\boxed{n \geq 1}} + \underbrace{\binom{5}{1}\binom{n}{2}}_{\boxed{n \geq 2}} + \underbrace{\binom{4}{1}\binom{n}{3}}_{\boxed{n \geq 3}}$$

$$\sum_{0}^{n-1} {}^{3}T_{i+1}^{[1]} = \sum_{0}^{n-1} \frac{(3i+1)(3i+2)}{2} \qquad [n = 1, 2, \cdots, \infty]$$

$$= \sum_{1}^{n} T_{1+3(i-1)} = n + (T_4 - T_1)\left[\frac{n(n-1)}{2} \right]$$

$$+ 3^2 \sum_{1}^{n-2} i\frac{(n-i)(n-(i+1))}{2}.$$

Substituting $a = 1$, $b = 9$, $c = 3^2 = 9$, in (5.2) we have

$$= n + 9\left[\frac{n(n-1)}{2} \right] + 9\sum_{1}^{n-2} i\frac{(n-i)(n-(i+1))}{2}$$

$$= n + 9\left[\frac{n(n-1)}{2!} \right] + 9\left[\frac{n(n-1)(n-2)}{3!} \right]$$

$$= \underbrace{\binom{n}{1}}_{\boxed{n \geq 1}} + \underbrace{\binom{9}{1}\binom{n}{2}}_{\boxed{n \geq 2}} + \underbrace{\binom{4}{1}\binom{n}{3}}_{\boxed{n \geq 3}}$$

$$\sum_{0}^{n-1} {}^{4}T_{i+1}^{[1]} = \sum_{0}^{n-1} \frac{(4i+1)(4i+2)}{2} \qquad [n = 1, 2, \cdots, \infty]$$

$$= \sum_{1}^{n} T_{1+4(i-1)} = n + (T_5 - T_1)\left[\frac{n(n-1)}{2} \right]$$

$$+ 4^2 \sum_{1}^{n-2} i\frac{(n-i)(n-(i+1))}{2}.$$

Substituting $a = 1$, $b = 14$, $c = 4^2 = 16$, in (5.2) we have

$$= n + 14\left[\frac{n(n-1)}{2} \right] + 16\sum_{1}^{n-2} i\frac{(n-i)(n-(i+1))}{2}$$

$$= n + 14\left[\frac{n(n-1)}{2!} \right] + 16\left[\frac{n(n-1)(n-2)}{3!} \right]$$

$$= \binom{n}{1} + \binom{14}{1}\binom{n}{2} + \binom{16}{1}\binom{n}{3}$$

$$\boxed{n \geq 1} \qquad \boxed{n \geq 2} \qquad \boxed{n \geq 3}$$

$- - - - - - -$

$$\sum_{0}^{n-1} {}^{\phi}\mathbf{T_{i+1}} = \sum_{0}^{n-1} \frac{(\phi i + 1)(\phi i + 2)}{2} [n = 1, 2, \cdots, \infty]$$

$$= \sum_{1}^{n} T_{1 + \phi(i-1)} = n + (T_{1+\phi} - T_1)\left[\frac{n(n-1)}{2}\right]$$

$$+ \phi^2 \sum_{1}^{n-2} i\frac{(n-i)(n-(i+1))}{2}.$$

Substituting $a = 1$, $b = \left[\frac{(\phi+1)(\phi+2)}{2} - 1\right]$, $c = \phi^2$, in (5.2) we have

$$= n\left[\frac{(\phi+1)(\phi+2)}{2} - 1\right]\left[\frac{n(n-1)}{2}\right] + \phi^2 \sum_{1}^{n-2} i\frac{(n-i)(n-(i+1))}{2}$$

$$= n + \left[\frac{(\phi+1)(\phi+2)}{2} - 1\right]\left[\frac{n(n-1)}{2!}\right] + \phi^2\left[\frac{n(n-1)(n-2)}{3!}\right]$$

$$= \binom{n}{1} + \left(\left[\frac{\frac{(\phi+1)(\phi+2)}{2} - 1}{1}\right]\right)\binom{n}{2} + \binom{\phi^2}{1}\binom{n}{3}$$

$$\boxed{n \geq 1} \qquad\qquad\qquad \boxed{n \geq 2} \quad \boxed{n \geq 3}$$

$$\sum_{0}^{n-1} {}^{1}\mathbf{T_{i+1}} = \sum_{0}^{n-1} \frac{(2i+1)(2i+2)}{2} [n = 1, 2, \cdots, \infty]$$

$$= \sum_{1}^{n} T_{1 + 2(i-1)} = n + (T_3 - T_1)\left[\frac{n(n-1)}{2}\right]$$

$$+ 2^2 \sum_{1}^{n-2} i\frac{(n-i)(n-(i+1))}{2}.$$

Substituting $a = 1$, $b = 5$, $c = 2^2 = 4$, in (5.2) we have

$$= n + 5\left[\frac{n(n-1)}{2}\right] + 4 \sum_{1}^{n-2} i\frac{(n-i)(n-(i+1))}{2}$$

$$= n + 5\left[\frac{n(n-1)}{2!}\right] + 4\left[\frac{n(n-1)(n-2)}{3!}\right]$$

$$= \binom{n}{1} + \binom{5}{1}\binom{n}{2} + \binom{4}{1}\binom{n}{3}$$
$$\boxed{n \geq 1} \qquad \boxed{n \geq 2} \qquad \boxed{n \geq 3}$$

$$\sum_{0}^{n-1} {}^{2}T_{i+1}^{[2]} = \sum_{0}^{n-1} \frac{(4i+1)(4i+2)}{2} \qquad [n = 1, 2, \cdots, \infty]$$

$$= \sum_{1}^{n} T_{1+4(i-1)} = n + (T_5 - T_1)\left[\frac{n(n-1)}{2}\right]$$

$$+ 4^2 \sum_{1}^{n-2} i\frac{(n-i)(n-(i+1))}{2}.$$

Substituting $a = 1$, $b = 14$, $c = 4^2 = 16$, in (5.2) we have

$$= n + 14\left[\frac{n(n-1)}{2}\right] + 16\sum_{1}^{n-2} i\frac{(n-i)(n-(i+1))}{2}$$

$$= n + 14\left[\frac{n(n-1)}{2!}\right] + 16\left[\frac{n(n-1)(n-2)}{3!}\right]$$

$$= \binom{n}{1} + \binom{14}{1}\binom{n}{2} + \binom{16}{1}\binom{n}{3}$$
$$\boxed{n \geq 1} \qquad \boxed{n \geq 2} \qquad \boxed{n \geq 3}$$

$$\sum_{0}^{n-1} {}^{3}T_{i+1}^{[2]} = \sum_{0}^{n-1} \frac{(6i+1)(6i+2)}{2} \qquad [n = 1, 2, \cdots, \infty]$$

$$= \sum_{1}^{n} T_{1+6(i-1)} = n + (T_7 - T_1)\left[\frac{n(n-1)}{2}\right]$$

$$+ 6^2 \sum_{1}^{n-2} i\frac{(n-i)(n-(i+1))}{2}.$$

Substituting $a = 1$, $b = 27$, $c = 6^2 = 36$, in (5.2) we have

$$= n + 27\left[\frac{n(n-1)}{2}\right] + 36\sum_{1}^{n-2} i\frac{(n-i)(n-(i+1))}{2}$$

$$= n + 27\left[\frac{n(n-1)}{2!}\right] + 36\left[\frac{n(n-1)(n-2)}{3!}\right]$$

$$= \binom{n}{1} + \binom{27}{1}\binom{n}{2} + \binom{36}{1}\binom{n}{3}$$
$$\boxed{n \geq 1} \qquad \boxed{n \geq 2} \qquad \boxed{n \geq 3}$$

$$\sum_{0}^{n-1}{}^{4}\mathbf{T}_{\mathbf{i}+\mathbf{1}}^{[\mathbf{2}]} = \sum_{0}^{n-1}\frac{(8i+1)(8i+2)}{2} \qquad [n=1,2,\cdots,\infty]$$

$$= \sum_{1}^{n} T_{1+8(i-1)} = n + (T_9 - T_1)\left[\frac{n(n-1)}{2}\right]$$

$$+ 8^2 \sum_{1}^{n-2} i\frac{(n-i)(n-(i+1))}{2}.$$

Substituting $a = 1$, $b = 44$, $c = 8^2 = 64$, in (5.2) we have

$$= n + 44\left[\frac{n(n-1)}{2}\right] + 64\sum_{1}^{n-2} i\frac{(n-i)(n-(i+1))}{2}$$

$$= n + 44\left[\frac{n(n-1)}{2!}\right] + 64\left[\frac{n(n-1)(n-2)}{3!}\right]$$

$$= \binom{n}{1} + \binom{44}{1}\binom{n}{2} + \binom{64}{1}\binom{n}{3}$$

$$\boxed{n \geq 1} \qquad \boxed{n \geq 2} \qquad \boxed{n \geq 3}$$

– – – – – – – –

$$\sum_{0}^{n-1}{}^{\phi}\mathbf{T}_{\mathbf{i}+\mathbf{1}}^{[\mathbf{2}]} = \sum_{0}^{n-1}\frac{(2\phi i+1)(2\phi i+2)}{2} \qquad [n=1,2,\cdots,\infty]$$

$$= \sum_{1}^{n} T_{1+2\phi(i-1)}$$

$$= n + (T_{1+2\phi} - T_1)\left[\frac{n(n-1)}{2}\right] + (2\phi)^2 \sum_{1}^{n-2} i\frac{(n-i)(n-(i+1))}{2}.$$

Substituting $a = 1$, $b = \left[\frac{(2\phi+1)(2\phi+2)}{2} - 1\right]$, $c = (2\phi)^2$, in (5.2) we have

$$= n + \left[\frac{(2\phi+1)(2\phi+2)}{2} - 1\right]\left[\frac{n(n-1)}{2}\right]$$

$$+ (2\phi)^2 \sum_{1}^{n-2} i\frac{(n-i)(n-(i+1))}{2}$$

$$= n + \left[\frac{(2\phi+1)(2\phi+2)}{2} - 1\right]\left[\frac{n(n-1)}{2!}\right] + (2\phi)^2\left[\frac{n(n-1)(n-2)}{3!}\right]$$

$$= \binom{n}{1} + \left(\left[\frac{\frac{(2\phi+1)(2\phi+2)}{2}-1}{1}\right]\right)\binom{n}{2} + \binom{(2\phi)^2}{1}\binom{n}{3}$$

$$\boxed{n \geq 1} \qquad\qquad\qquad\qquad\qquad\qquad \boxed{n \geq 2} \quad \boxed{n \geq 3}$$

$$\sum_{0}^{n-1} {}^{1}\mathbf{T_{i+1}}_{2}^{[3]} = \sum_{0}^{n-1} \frac{(3i+1)(3i+2)}{2} \qquad [n = 1, 2, \cdots, \infty]$$

$$= \sum_{1}^{n} T_{1+3(i-1)} = n + (T_4 - T_1)\left[\frac{n(n-1)}{2}\right]$$

$$+ 3^2 \sum_{1}^{n-2} i\frac{(n-i)(n-(i+1))}{2}$$

Substituting $a = 1$, $b = 9$, $c = 3^2 = 9$, in (5.2) we have

$$= n + 9\left[\frac{n(n-1)}{2}\right] + 9\sum_{1}^{n-2} i\frac{(n-i)(n-(i+1))}{2}$$

$$= n + 9\left[\frac{n(n-1)}{2!}\right] + 9\left[\frac{n(n-1)(n-2)}{3!}\right]$$

$$= \binom{n}{1} + \binom{9}{1}\binom{n}{2} + \binom{9}{1}\binom{n}{3}$$

$$\boxed{n \geq 1} \qquad \boxed{n \geq 2} \qquad \boxed{n \geq 3}$$

$$\sum_{0}^{n-1} {}^{2}\mathbf{T_{i+1}}^{[3]} = \sum_{0}^{n-1} \frac{(6i+1)(6i+2)}{2} \qquad [n = 1, 2, \cdots, \infty]$$

$$= \sum_{1}^{n} T_{1+6(i-1)} = n + (T_7 - T_1)\left[\frac{n(n-1)}{2}\right]$$

$$+ 6^2 \sum_{1}^{n-2} i\frac{(n-i)(n-(i+1))}{2}.$$

Substituting $a = 1$, $b = 27$, $c = 6^2 = 36$, in (5.2) we have

$$= n + 27\left[\frac{n(n-1)}{2}\right] + 36\sum_{1}^{n-2} i\frac{(n-i)(n-(i+1))}{2}$$

$$= n + 27\left[\frac{n(n-1)}{2!}\right] + 36\left[\frac{n(n-1)(n-2)}{3!}\right]$$

$$= \binom{n}{1} + \binom{27}{1}\binom{n}{2} + \binom{36}{1}\binom{n}{3}$$

$$\boxed{n \geq 1} \qquad \boxed{n \geq 2} \qquad \boxed{n \geq 3}$$

$$\sum_0^{n-1} {}^3\mathbf{T_{i+1}}^{[3]} = \sum_0^{n-1} \frac{(9i+1)(9i+2)}{2} \qquad [n = 1, 2, \cdots, \infty]$$

$$= \sum_1^n T_{1+9(i-1)} = n + (T_{10} - T_1)\left[\frac{n(n-1)}{2}\right]$$

$$+ 9^2 \sum_1^{n-2} i\frac{(n-i)(n-(i+1))}{2}.$$

Substituting $a = 1$, $b = 54$, $c = 9^2 = 81$, in (5.2) we have

$$= n + 54\left[\frac{n(n-1)}{2}\right] + 81 \sum_1^{n-2} i\frac{(n-i)(n-(i+1))}{2}$$

$$= n + 54\left[\frac{n(n-1)}{2!}\right] + 81\left[\frac{n(n-1)(n-2)}{3!}\right]$$

$$= \binom{n}{1} + \binom{54}{1}\binom{n}{2} + \binom{81}{1}\binom{n}{3}$$

$$\boxed{n \geq 1} \qquad \boxed{n \geq 2} \qquad \boxed{n \geq 3}$$

$$\sum_0^{n-1} {}^4\mathbf{T_{i+1}}^{[3]} = \sum_0^{n-1} \frac{(12i+1)(12i+2)}{2} \qquad [n = 1, 2, \cdots, \infty]$$

$$= \sum_1^n T_{1+12(i-1)} = n + (T_{13} - T_1)\left[\frac{n(n-1)}{2}\right]$$

$$+ (12)^2 \sum_1^{n-2} i\frac{(n-i)(n-(i+1))}{2}.$$

Substituting $a = 1$, $b = 90$, $c = (12)^2 = 144$, in (5.2) we have

$$= n + 90\left[\frac{n(n-1)}{2}\right] + 144 \sum_1^{n-2} i\frac{(n-i)(n-(i+1))}{2}$$

$$= n + 90\left[\frac{n(n-1)}{2!}\right] + 144\left[\frac{n(n-1)(n-2)}{3!}\right]$$

$$= \binom{n}{1} + \binom{90}{1}\binom{n}{2} + \binom{144}{1}\binom{n}{3}$$

$$\boxed{n \geq 1} \qquad \boxed{n \geq 2} \qquad \boxed{n \geq 3}$$

$- - - - - - - - - -$

$$\sum_{0}^{n-1} {}^{\phi}\mathbf{T_{i+1}}^{[3]} = \sum_{0}^{n-1} \frac{(3\phi i + 1)(3\phi i + 2)}{2} \qquad [n = 1, 2, \cdots, \infty]$$

$$= \sum_{1}^{n} T_{1+3\phi(i-1)} = n + (T_{1+3\phi} - T_1) \left[\frac{n(n-1)}{2} \right]$$

$$+ (3\phi)^2 \sum_{1}^{n-2} i \frac{(n-i)(n-(i+1))}{2}.$$

Substituting $a = 1$, $b = \left[\frac{(3\phi + 1)(3\phi + 2)}{2} - 1 \right]$, $c = (3\phi)^2$, in (5.2) we have

$$= n + \left[\frac{(3\phi + 1)(3\phi + 2)}{2} - 1 \right] \left[\frac{n(n-1)}{2} \right]$$

$$+ (3\phi)^2 \sum_{1}^{n-2} i \frac{(n-i)(n-(i+1))}{2}$$

$$= n + \left[\frac{(3\phi + 1)(3\phi + 2)}{2} - 1 \right] \left[\frac{n(n-1)}{2!} \right] + (3\phi)^2 \left[\frac{n(n-1)(n-2)}{3!} \right]$$

$$= \underset{\boxed{n \geq 1}}{\binom{n}{1}} + \underset{\boxed{n \geq 2}}{\left(\left[\frac{(3\phi + 1)(3\phi + 2)}{2} - 1 \right]_{1} \right) \binom{n}{2}} + \underset{\boxed{n \geq 3}}{\binom{(3\phi)^2}{1} \binom{n}{3}}$$

$$\sum_{0}^{n-1} {}^{1}\mathbf{T_{i+1}}^{[4]} = \sum_{0}^{n-1} \frac{(4i + 1)(4i + 2)}{2} \qquad [n = 1, 2, \cdots, \infty]$$

$$= \sum_{1}^{n} T_{1+4(i-1)} = n + (T_5 - T_1) \left[\frac{n(n-1)}{2} \right]$$

$$+ 4^2 \sum_{1}^{n-2} i \frac{(n-i)(n-(i+1))}{2}.$$

Substituting $a = 1$, $b = 14$, $c = 4^2 = 16$, in (5.2) we have

$$= n + 14 \left[\frac{n(n-1)}{2} \right] + 16 \sum_{1}^{n-2} i \frac{(n-i)(n-(i+1))}{2}$$

$$= n + 14 \left[\frac{n(n-1)}{2!} \right] + 16 \left[\frac{n(n-1)(n-2)}{3!} \right]$$

$$= \binom{n}{1} + \binom{14}{1}\binom{n}{2} + \binom{16}{1}\binom{n}{3}$$
$$\boxed{n \geq 1} \qquad \boxed{n \geq 2} \qquad \boxed{n \geq 3}$$

$$\sum_{0}^{n-1} {}^{2}\mathbf{T}_{\mathbf{i}+\mathbf{1}}^{[4]} = \sum_{0}^{n-1} \frac{(8i+1)(8i+2)}{2} \qquad [n = 1, 2, \cdots, \infty]$$

$$= \sum_{1}^{n} T_{1+8(i-1)} = n + (T_9 - T_1)\left[\frac{n(n-1)}{2}\right]$$

$$+ 8^2 \sum_{1}^{n-2} i\frac{(n-i)(n-(i+1))}{2}.$$

Substituting $a = 1$, $b = 44$, $c = (8)^2 = 64$, in (5.2) we have

$$= n + 44\left[\frac{n(n-1)}{2}\right] + 64\sum_{1}^{n-2} i\frac{(n-i)(n-(i+1))}{2}$$

$$= n + 44\left[\frac{n(n-1)}{2!}\right] + 64\left[\frac{n(n-1)(n-2)}{3!}\right]$$

$$= \binom{n}{1} + \binom{44}{1}\binom{n}{2} + \binom{64}{1}\binom{n}{3}$$
$$\boxed{n \geq 1} \qquad \boxed{n \geq 2} \qquad \boxed{n \geq 3}$$

$$\sum_{0}^{n-1} {}^{3}\mathbf{T}_{\mathbf{i}+\mathbf{1}}^{[4]} = \sum_{0}^{n-1} \frac{(12i+1)(12i+2)}{2} \qquad [n = 1, 2, \cdots, \infty]$$

$$= \sum_{1}^{n} T_{1+12(i-1)} = n + (T_{13} - T_1)\left[\frac{n(n-1)}{2}\right]$$

$$+ (12)^2 \sum_{1}^{n-2} i\frac{(n-i)(n-(i+1))}{2}$$

Substituting $a = 1$, $b = 90$, $c = (12)^2 = 144$, in (5.2) we have

$$= n + 90\left[\frac{n(n-1)}{2}\right] + 144\sum_{1}^{n-2} i\frac{(n-i)(n-(i+1))}{2}$$

$$= n + 90\left[\frac{n(n-1)}{2!}\right] + 144\left[\frac{n(n-1)(n-2)}{3!}\right]$$

$$= \binom{n}{1} + \binom{90}{1}\binom{n}{2} + \binom{144}{1}\binom{n}{3}$$
$$\boxed{n \geq 1} \qquad \boxed{n \geq 2} \qquad \boxed{n \geq 3}$$

$$\sum_{0}^{n-1} {}^4\mathbf{T_{i+1}^{[4]}} = \sum_{0}^{n-1} \frac{(16i+1)(16i+2)}{2} \qquad [n = 1, 2, \cdots, \infty]$$

$$= \sum_{1}^{n} T_{1+16(i-1)} = n + (T_{17} - T_1)\left[\frac{n(n-1)}{2}\right]$$

$$+ (16)^2 \sum_{1}^{n-2} i\frac{(n-i)(n-(i+1))}{2}.$$

Substituting $a = 1$, $b = 152$, $c = (16)^2 = 256$, in (5.2) we have

$$= n + 152\left[\frac{n(n-1)}{2}\right] + 256 \sum_{1}^{n-2} i\frac{(n-i)(n-(i+1))}{2}$$

$$= n + 152\left[\frac{n(n-1)}{2!}\right] + 256 \left[\frac{n(n-1)(n-2)}{3!}\right]$$

$$= \binom{n}{1} + \binom{152}{1}\binom{n}{2} + \binom{256}{1}\binom{n}{3}$$

$$\boxed{n \geq 1} \qquad \boxed{n \geq 2} \qquad \boxed{n \geq 3}$$

$$- - - - - - - - -$$

$$\sum_{0}^{n-1} {}^\phi\mathbf{T_{i+1}^{[4]}} = \sum_{0}^{n-1} \frac{(\phi 4i+1)(\phi 4i+2)}{2} \qquad [n = 1, 2, \cdots, \infty]$$

$$= \sum_{1}^{n} T_{1+4\phi(i-1)} = n + (T_{1+4\phi} - T_1)\left[\frac{n(n-1)}{2}\right]$$

$$+ (4\phi)^2 \sum_{1}^{n-2} i\frac{(n-i)(n-(i+1))}{2}.$$

Substituting $a = 1$, $b = \left[\frac{(4\phi+1)(4\phi+2)}{2} - 1\right]$, $c = (4\phi)^2$, in (5.2) we have

$$= n + \left[\frac{(4\phi+1)(4\phi+2)}{2} - 1\right]\left[\frac{n(n-1)}{2}\right]$$

$$+ (4\phi)^2 \sum_{1}^{n-2} i\frac{(n-i)(n-(i+1))}{2}$$

$$= n + \left[\frac{(4\phi+1)(4\phi+2)}{2} - 1\right]\left[\frac{n(n-1)}{2!}\right] + (4\phi)^2\left[\frac{n(n-1)(n-2)}{3!}\right]$$

$$= \binom{n}{1} + \left(\left[\frac{(4\phi+1)(4\phi+2)}{2} - 1 \right] \atop 1 \right) \binom{n}{2} + \binom{(4\phi)^2}{1} \binom{n}{3}$$

$$\boxed{n \geq 1} \qquad\qquad\qquad\qquad \boxed{n \geq 2} \qquad \boxed{n \geq 3}$$

– – – – – – – –

$$\sum_{0}^{n-1} {}^{1}\mathbf{T_{i+1}^{[R]}} = \sum_{0}^{n-1} \frac{(Ri+1)(Ri+2)}{2} \qquad [n = 1, 2, \cdots, \infty]$$

$$= \sum_{1}^{n} T_{1+R(i-1)} = n + (T_{1+R} - T_1) \left[\frac{n(n-1)}{2} \right]$$

$$+ R^2 \sum_{1}^{n-2} i \frac{(n-i)(n-(i+1))}{2}$$

Substituting $a = 1$, $b = \left[\frac{(R+1)(R+2)}{2} - 1 \right]$, $c = R^2$, in (5.2) we have

$$= n + \left[\frac{(R+1)(R+2)}{2} - 1 \right] \left[\frac{n(n-1)}{2} \right] + R^2 \sum_{1}^{n-2} i \frac{(n-i)(n-(i+1))}{2}$$

$$= n + \left[\frac{(R+1)(R+2)}{2} - 1 \right] \left[\frac{n(n-1)}{2!} \right] + R^2 \left[\frac{n(n-1)(n-2)}{3!} \right]$$

$$= \binom{n}{1} + \left(\left[\frac{(R+1)(R+2)}{2} - 1 \right] \atop 1 \right) \binom{n}{2} + \binom{R^2}{1} \binom{n}{3}$$

$$\boxed{n \geq 1} \qquad\qquad\qquad\qquad \boxed{n \geq 2} \qquad \boxed{n \geq 3}$$

$$\sum_{0}^{n-1} {}^{2}\mathbf{T_{i+1}^{[R]}} = \sum_{0}^{n-1} \frac{(2Ri+1)(2Ri+2)}{2} \qquad [n = 1, 2, \cdots, \infty]$$

$$= \sum_{1}^{n} T_{1+2R(i-1)} = n + (T_{1+2R} - T_1) \left[\frac{n(n-1)}{2} \right]$$

$$+ (2R)^2 \sum_{1}^{n-2} i \frac{(n-i)(n-(i+1))}{2}$$

Substituting $a = 1$, $b = \left[\frac{(2R+1)(2R+2)}{2} - 1 \right]$, $c = (2R)^2$, in (5.2) we have

$$= n + \left[\frac{(2R+1)(2R+2)}{2} - 1 \right] \left[\frac{n(n-1)}{2} \right] + (2R)^2 \sum_{1}^{n-2} i \frac{(n-i)(n-(i+1))}{2}$$

$$= n + \left[\frac{(2R+1)(2R+2)}{2} - 1\right]\left[\frac{n(n-1)}{2!}\right] + (2R)^2\left[\frac{n(n-1)(n-2)}{3!}\right]$$

$$= \underbrace{\binom{n}{1}}_{\boxed{n \geq 1}} + \underbrace{\left(\left[\frac{\frac{(2R+1)(2R+2)}{2} - 1}{1}\right]\right)\binom{n}{2}}_{\boxed{n \geq 2}} + \underbrace{\binom{(2R)^2}{1}\binom{n}{3}}_{\boxed{n \geq 3}}$$

$$\sum_0^{n-1} {}^3\mathbf{T}_{\mathbf{i}+\mathbf{1}}^{[\mathbf{R}]} = \sum_0^{n-1} \frac{(3Ri+1)(3Ri+2)}{2} \qquad [n = 1, 2, \cdots, \infty]$$

$$= \sum_1^n T_{1+3R(i-1)} = n + (T_{1+3R} - T_1)\left[\frac{n(n-1)}{2}\right]$$

$$+ (3R)^2 \sum_1^{n-2} i\frac{(n-i)(n-(i+1))}{2}.$$

Substituting $a = 1$, $b = \left[\frac{(3R+1)(3R+2)}{2} - 1\right]$, $c = (3R)^2$, in (5.2) we have

$$= n + \left[\frac{(3R+1)(3R+2)}{2} - 1\right]\left[\frac{n(n-1)}{2}\right]$$

$$+ (3R)^2 \sum_1^{n-2} i\frac{(n-i)(n-(i+1))}{2}$$

$$= n + \left[\frac{(3R+1)(3R+2)}{2} - 1\right]\left[\frac{n(n-1)}{2!}\right] + (3R)^2\left[\frac{n(n-1)(n-2)}{3!}\right]$$

$$= \underbrace{\binom{n}{1}}_{\boxed{n \geq 1}} + \underbrace{\left(\left[\frac{\frac{(3R+1)(3R+2)}{2} - 1}{1}\right]\right)\binom{n}{2}}_{\boxed{n \geq 2}} + \underbrace{\binom{(3R)^2}{1}\binom{n}{3}}_{\boxed{n \geq 3}}$$

$$\sum_0^{n-1} {}^4\mathbf{T}_{\mathbf{i}+\mathbf{1}}^{[\mathbf{R}]} = \sum_0^{n-1} \frac{(4Ri+1)(4Ri+2)}{2} \qquad [n = 1, 2, \cdots, \infty]$$

$$= \sum_1^n T_{1+4R(i-1)} = n + (T_{1+4R} - T_1)\left[\frac{n(n-1)}{2}\right]$$

$$+ (4R)^2 \sum_1^{n-2} i\frac{(n-i)(n-(i+1))}{2}$$

Substituting $a = 1$, $b = \left[\dfrac{(4R+1)(4R+2)}{2} - 1\right]$, $c = (4R)^2$, in (5.2) we have

$$= n + \left[\frac{(4R+1)(4R+2)}{2} - 1\right]\left[\frac{n(n-1)}{2}\right]$$

$$+ (4R)^2 \sum_{1}^{n-2} i\frac{(n-i)(n-(i+1))}{2}$$

$$= n + \left[\frac{(4R+1)(4R+2)}{2} - 1\right]\left[\frac{n(n-1)}{2!}\right] + (4R)^2\left[\frac{n(n-1)(n-2)}{3!}\right]$$

$$= \underset{n \geq 1}{\binom{n}{1}} + \left(\left[\frac{\frac{(4R+1)(4R+2)}{2} - 1}{1}\right]\right)\underset{n \geq 2}{\binom{n}{2}} + \underset{n \geq 3}{\binom{(4R)^2}{1}\binom{n}{3}}$$

$- - - - - - -$

$$\sum_{0}^{n-1} {}^{\phi}\mathbf{T_{i+1}^{[R]}} = \sum_{0}^{n-1} \frac{(\phi R i + 1)(\phi R i + 2)}{2} \qquad [n = 1, 2, \cdots, \infty]$$

$$= \sum_{1}^{n} T_{1+\phi R(i-1)} = n + (T_{1+\phi R} - T_1)\left[\frac{n(n-1)}{2}\right]$$

$$+ (\phi R)^2 \sum_{1}^{n-2} i\frac{(n-i)(n-(i+1))}{2}$$

Substituting $a = 1$, $b = \left[\dfrac{(\phi R+1)(\phi R+2)}{2} - 1\right]$, $c = (\phi R)^2$, in (5.2) we have

$$= n + \left[\frac{(\phi R+1)(\phi R+2)}{2} - 1\right]\left[\frac{n(n-1)}{2}\right]$$

$$+ (\phi R)^2 \sum_{1}^{n-2} i\frac{(n-i)(n-(i+1))}{2}$$

$$= n + \left[\frac{(\phi R+1)(\phi R+2)}{2} - 1\right]\left[\frac{n(n-1)}{2!}\right] + (\phi R)^2\left[\frac{n(n-1)(n-2)}{3!}\right]$$

$$= \underset{n \geq 1}{\binom{n}{1}} + \left(\left[\frac{\frac{(\phi R+1)(\phi R+2)}{2} - 1}{1}\right]\right)\underset{n \geq 2}{\binom{n}{2}} + \underset{n \geq 3}{\binom{(\phi R)^2}{1}\binom{n}{3}}.$$

6. Some Interesting Properties of Triangular Numbers

We have proved Theorem 1.1 that
If $T = \{T_0, T_1, T_2. T_3, \cdots, T_n, \cdots, \infty\}[T_0 = 0]$ are Triangular Numbers then $(2\xi + 1)^2 T_n + T_\xi [\xi \geq 0, n \geq 0]$ are also Triangular Numbers.

We can utilize this theorem to generate Infinite Inductive Theorems as follows. The self-evident (though a bit cumbersome) inductive steps of the Proofs are omitted.

Theorem 6.1. *There Exist Infinite Triangular Numbers That Are the sum of three Triangular Numbers.*

Proof. Substituting $n = 1$ in "$(T_{2\xi} + T_{2\xi+1})T_n + T_\xi = T_{(3\xi n+1)}$"   [1.1]

we have $T_{2\xi} + T_{2\xi+1} + T_\xi = T_{(3\xi+1)} \cdot [\xi \geq 1]$ proving the theorem.   □

Theorem 6.2. *There Exist Infinite Triangular Numbers That Are the sum of Five Triangular Numbers.*

Proof. $((2\xi + 1)^2)^2 + \mathbf{T_{(3\xi+1)}} = (T_{2\xi} + T_{2\xi+1})^2 + T_{(3\xi+1)}$

$$= T_{(2\xi+1)^2} + T_{[(2\xi+1)^2+1]} + T_{2\xi} + T_{2\xi+1} + T_\xi$$

$$= T_{[(2\xi+1)^2+(3\xi+1)]} \cdot [\xi \geq 1] \ \text{proving the theorem.} \quad □$$

Theorem 6.3. *There Exist Infinite Triangular Numbers That Are the sum of Seven Triangular Numbers.*

Proof. $(((2\xi + 1)^2)^2)^2 + \mathbf{T_{[(2\xi+1)^2+(3\xi+1)]}} =$

$T_{[((2\xi+1)^2)^2]} + T_{[((2\xi+1)^2)^2+1]} +$
$T_{(2\xi+1)^2} + T_{[(2\xi+1)^2+1]} + T_{2\xi} + T_{2\xi+1} + T_\xi$
$$= T_{[((2\xi+1)^2)^2+(2\xi+1)^2+(3\xi+1)]}. \quad [\xi \geq 1]$$

proving the theorem.

□

Theorem 6.4. *There Exist Infinite Triangular Numbers That Are the sum of Nine Triangular Numbers.*

Proof. $((((2\xi + 1)^2)^2)^2)^2 + \mathbf{T_{[((2\xi+1)^2)^2+(2\xi+1)^2+(3\xi+1)]}} =$

$T_{[(((2\xi+1)^2)^2)^2]} + T_{[(((2\xi+1)^2)^2)^2+1]} + T_{[((2\xi+1)^2)^2]} +$
$T_{[((2\xi+1)^2)^2]+1]} + T_{(2\xi+1)^2} + T_{[(2\xi+1)^2+1]} + T_{2\xi} + T_{2\xi+1} + T_\xi$
$= T_{[(((2\xi+1)^2)^2)^2+((2\xi+1)^2)^2+(2\xi+1)^2+(3\xi+1)]} \cdot \quad [\xi \geq 1]$

proving the theorem.

□

Theorem 6.τ $[\tau \geq 1]$. *There Exist Infinite Triangular Numbers That Are the sum of $(2\tau + 1)$, $[\tau \geq 1]$ Triangular Numbers.*

We continue the Inductive Proof for $[\tau \geq 3]$

$(((((2\xi + 1)^2)^2)^2)^{\tau \ \text{times}} +$

$T_{[((((2\xi+1)^2)^2))))^{(\tau-2)\text{times}}\ldots+((2\xi+1)^2)^2+(2\xi+1)^2+(3\xi+1)]} =$

$$T_{[((((2\xi+1)^2)^2)^2)(\tau-1)\text{times}]} + T_{[(((((2\xi+1)^2)^2)^2)(\tau-1)\text{times}+1]}+$$

$$T_{[((((2\xi+1)^2)^2)^2)(\tau-2)\text{times}]} + T_{[(((((2\xi+1)^2)^2)^2)(\tau-2)\text{times}+1]}+$$

$$T_{[((((2\xi+1)^2)^2)^2)(\tau-3)\text{times}]} + T_{[(((((2\xi+1)^2)^2)^2)(\tau-3)\text{times}+1]}+$$

$$-\ -\ -\ -\ -\ -\ -$$

$$T_{[(((2\xi+1)^2)^2)]} + T_{[(((2\xi+1)^2)^2)+1]}+$$

$$T_{(2\xi+1)^2} + T_{[(2\xi+1)^2+1]} + T_{2\xi} + T_{2\xi+1} + T_{\xi}$$

$$= T_{[(((((2\xi+1)^2)^2)))(\tau-1)\text{times}+\cdots+((2\xi+1)^2)^2} + (2\xi+1)^2 + (3\xi+1)]\cdot$$

$$[\xi \geq 1][\tau \geq 3]$$

Simplifying

$$(2\xi+1)^{2^{\tau}} + T_{[(2\xi+1)^{2(\tau-2)}+\cdots+(2\xi+1)^4+(2\xi+1)^2+(3\xi+1)]} =$$

$$T_{[(2\xi+1)^{2(\tau-1)}]} + T_{[(2\xi+1)^{2(\tau-1)}+1]}+$$

$$T_{[(2\xi+1)^{2(\tau-2)}]} + T_{[(2\xi+1)^{2(\tau-2)}+1]}+$$

$$T_{[(2\xi+1)^{2(\tau-3)}]} + T_{[(2\xi+1)^{2(\tau-3)}+1]}+$$

$$-\ -\ -\ -\ -\ -\ -$$

$$T_{[(2\xi+1)^4]} + T_{[(2\xi+1)^4+1]} + T_{(2\xi+1)^2} + T_{[(2\xi+1)^2+1]}$$

$$+T_{2\xi} + T_{2\xi+1} + T_{\xi}$$

$$= T_{[(2\xi+1)^{2(\tau-1)}+\cdots+(2\xi+1)^4+(2\xi+1)^2+(3\xi+1)]} \cdots \quad [\xi \geq 1][\tau \geq 3]$$

proving the theorems.

We have Proved Theorem 1.5 that

If $T = \{T_0, T_1, T_2, T_3, \cdots T_n, \cdots, \cdots \infty\}[T_0 = 0]$ are Triangular Numbers then $(2\lambda)^2 T_n + T_\lambda + \lambda n [\lambda \geq 1, n \geq 0]$ are also Triangular Numbers.

We can utilise this theorem to generate Infinite Inductive Theorems as follows. The self-evident (though a bit cumbersome) inductive steps of the Proofs are omitted.

Theorem 6.5. *There Exist Infinite Triangular Numbers That Are the sum of three Triangular Numbers and a specified constant $\lambda = 1, 2, 3, \cdots \infty$ or Equivalently.*

For each $\lambda = 1, 2, 3, \cdots, \infty$ There Exist Three Triangular Numbers, the Sum of which and λ is yet another Triangular Number.

Proof. Substituting $n = 1$ in "$(T_{2\lambda-1} + T_{2\lambda})T_n + T_\lambda + \lambda n = T_{3\lambda n}$" [1.2] we have $T_{2\lambda-1} + T_{2\lambda} + T_\lambda + \lambda = T_{3\lambda}$ $[\lambda \geq 1]$ proving the theorem.

When λ is a Triangular Number we have the Infinite cases when we have Triangular Numbers that are the Sum of Four Triangular Numbers

$$\mathbf{T_{[n(n+1)-1]}} + \mathbf{T_{n(n+1)}} + \mathbf{T_{\frac{[n(n+1)]}{2}}} + \mathbf{T_n} = T_{\frac{3n(n+1)}{2}} \quad [n \geq 1]$$

proving the theorem.   □

Theorem 6.6. *There Exist Infinite Triangular Numbers That Are the sum of Five Triangular Numbers and a specified constant* $\lambda = 1, 2, 3, \cdots, \infty$ *or Equivalently.*

For each $\lambda = 1, 2, 3, \cdots, \infty$ *There Exist Five Triangular Numbers, the Sum of which and* λ *is yet another Triangular Number.*

Proof. $((2\lambda)^2)^2 + \mathbf{T_{3\lambda}} = (T_{2\lambda-1} + T_{2\lambda})^2 + T_{3\lambda}$

$$= T_{(2\lambda)^2} + T_{[(2\lambda)^2+1]} + T_{2\lambda-1} + T_{2\lambda} + T_{\lambda} + \lambda$$

$$= T_{[(2\lambda)^2+(3\lambda)]} \cdot \qquad [\lambda \geq 1] \text{ proving the theorem.}$$

When λ is a Triangular Number we have the Infinite cases when we have Triangular Numbers that are the Sum of Six Triangular Numbers

$$T_{[(n(n+1))^2]} + T_{[(n(n+1))^2+1]} + T_{[n(n+1)-1]} + T_{n(n+1)} + T_{\frac{[n(n+1)]}{2}} + T_n$$

$$= T_{[(n(n+1))^2+\frac{3n(n+1)}{2}]} \cdot \qquad [n \geq 1]$$

proving the theorem. $\qquad\qquad\qquad\qquad\qquad\qquad\qquad\qquad\qquad\qquad\qquad\qquad\quad \square$

Theorem 6.7. *There Exist Infinite Triangular Numbers That Are the sum of Seven Triangular Numbers and a specified constant* $\lambda = 1, 2, 3, \cdots, \infty$ *or Equivalently.*

For each $\lambda = 1, 2, 3, \cdots, \infty$ *There Exist Seven Triangular Numbers, the Sum of which and* λ *is yet another Triangular Number.*

Proof.

$$(((2\lambda)^2)^2)^2 + \mathbf{T_{[(2\lambda)^2+(3\lambda)]}} = T_{[((2\lambda)^2)^2]} + T_{[((2\lambda)^2)^2+1]}$$

$$+ T_{(2\lambda)^2} + T_{[(2\lambda)^2+1]} + T_{2\lambda-1} + T_{2\lambda} + T_{\lambda} + \lambda$$

$$= T_{[((2\lambda)^2)^2+(2\lambda)^2+(3\lambda)]} \cdot \quad [\lambda \geq 1]$$

proving the theorem.

When λ is a Triangular Number we have the Infinite cases when we have Triangular Numbers that are the Sum of Eight Triangular Numbers

$$T_{[((n(n+1))^2)^2]} + T_{[((n(n+1))^2)^2+1]} + T_{[(n(n+1))^2]} +$$

$$T_{[(n(n+1))^2+1]} + T_{[(n(n+1))-1]} + T_{(n(n+1))} + T_{\frac{[n(n+1)]}{2}} + T_n$$

$$= T_{[((n(n+1))^2)^2+(n(n+1))^2+\frac{3n(n+1)}{2}]} \cdot \quad [n \geq 1]$$

proving the theorem. $\qquad\qquad\qquad\qquad\qquad\qquad\qquad\qquad\qquad\qquad\qquad\qquad\quad \square$

Theorem 6.8. *There Exist Infinite Triangular Numbers That Are the sum of Nine Triangular Numbers and a specified constant* $\lambda = 1, 2, 3, \cdots, \infty$ *or Equivalently.*

For each $\lambda = 1, 2, 3, \cdots, \infty$ *There Exist Nine Triangular Numbers, the Sum of which and* λ *is yet another Triangular Number.*

Proof. $((((2\lambda)^2)^2)^2)^2 + \mathbf{T}_{[((2\lambda)^2)^2+(2\lambda)^2+(3\lambda)]} =$

$T_{[(((2\lambda)^2)^2)^2]} + T_{[(((2\lambda)^2)^2)^2+1]} + T_{[((2\lambda)^2)^2]} + T_{[((2\lambda)^2)^2+1]} +$

$T_{(2\lambda)^2} + T_{[(2\lambda)^2+1]} + T_{2\lambda-1} + T_{2\lambda} + T_{\lambda} + \lambda$

$= T_{[(((2\lambda)^2)^2)^2+((2\lambda)^2)^2+(2\lambda)^2+(3\lambda)]}. \quad [\lambda \geq 1]$

proving the theorem.

When λ is a Triangular Number we have the Infinite cases when we have Triangular Numbers that are the Sum of Ten Triangular Numbers

$T_{[(((n(n+1))^2)^2)^2]} + T_{[(((n(n+1))^2)^2)^2+1]}$

$T_{[((n(n+1))^2)^2]} + T_{[((n(n+1))^2)^2+1]}$

$T_{[(n(n+1))^2]} + T_{[(n(n+1))^2+1]} + T_{[n(n+1)-1]} + T_{n(n+1)} + T_{\frac{[n(n+1)]}{2}} + T_n$

$= T_{[(((n(n+1))^2)^2)^2+((n(n+1))^2)^2+(n(n+1))^2+\frac{3n(n+1)}{2}]}. \quad [n \geq 1]$

proving the theorem. $\square$

- - - - - -

Theorem 6. $\psi[\psi \geq 1]$. *There Exist Infinite Triangular Numbers That Are the sum of* $(2\psi + 1)$ $[\psi \geq 1]$ *Triangular Numbers and a specified constant* $\lambda = 1, 2, 3, \cdots, \infty$ *Or Equivalently.*

For each $\lambda = 1, 2, 3, \cdots, \infty$ *There Exist* $(2\psi + 1)[\psi \geq 1]$ *Triangular Numbers, the Sum of which and* λ *is yet another Triangular Number.*

We continue the Inductive Proof for $[\psi \geq 3]$

$((((2\lambda)^2)^2)^2))))^{\psi \text{ times}} +$

$T_{[(((2\lambda)^2)^2))))(\psi-2) \text{ times}\ldots+((2\lambda)^2)^2(2\lambda)^2+(3\lambda)]} =$

$T_{[(((((2\lambda)^2)^2)^2))(\psi-1) \text{ times}]} + T_{[(((((2\lambda)^2)^2)^2))(\psi-1) \text{ times}+1]} +$

$T_{[(((((2\lambda)^2)^2)^2))(\psi-2) \text{ times}]} + T_{[(((((2\lambda)^2)^2)^2))(\psi-2) \text{ times}+1]} +$

$T_{[(((((2\lambda)^2)^2)^2))(\psi-3) \text{ times}]} + T_{[(((((2\lambda)^2)^2)^2))(\psi-3) \text{ times}+1]} +$

- - - - - - - -

$T_{[((2\lambda)^2)^2]} + T_{[((2\lambda)^2)^2+1]} + T_{(2\lambda)^2} + T_{[(2\lambda)^2+1]} + T_{2\lambda-1} + T_{2\lambda} + T_{\lambda} + \lambda$

$= T_{[(((((2\lambda)^2)^2)^2)(\psi-1) \text{ times}+\cdots+((2\lambda)^2)^2+(2\lambda)^2+(3\lambda)]}. \quad [\lambda \geq 1] \quad [\psi \geq 3]$

Simplifying

$(2\lambda)^{2^{\psi}} + T_{[(2\lambda)^{2^{\psi-2}}+\cdots+(2\lambda)^4+(2\lambda)^2+(3\lambda)]} =$

$T_{[(2\lambda)^{2^{(\psi-1)}}]} + T_{[(2\lambda)^{2^{(\psi-1)}}+1]} +$

$T_{[(2\lambda)^{2^{(\psi-2)}}]} + T_{[(2\lambda)^{2^{(\psi-2)}}+1]} +$

$T_{[(2\lambda)^{2^{(\psi-3)}}]} + T_{[(2\lambda)^{2^{(\psi-3)}}+1]} +$

- - - - - - - - -

$$T_{[(2\lambda)^4]} + T_{[(2\lambda)^4+1]} + T_{(2\lambda)^2} + T_{[(2\lambda)^2+1]} + T_{2\lambda-1} + T_{2\lambda} + T_\lambda + \lambda$$

$$= T_{[(2\lambda)^{2(\psi-1)}+\cdots+(2\lambda)^4+(2\lambda)^2+(3\lambda)]} \cdot \qquad\qquad [\lambda \geq 1][\psi \geq 3]$$

proving the theorems.

When λ is a Triangular Number we have the Infinite cases when we have Triangular Numbers that are the Sum of $(2\psi + 2) = 2(\psi + 1)[\psi \geq 1]$ Triangular Numbers.

We continue the Inductive Proof for $[\psi \geq 3]$

$$T_{[(n(n+1))^{2(\psi-1)}]} + T_{[(n(n+1))^{2(\psi-1)}+1]}+$$

$$T_{[(n(n+1))^{2(\psi-2)}]} + T_{[(n(n+1))^{2(\psi-2)}+1]}+$$

$$T_{[(n(n+1))^{2(\psi-3)}]} + T_{[(n(n+1))^{2(\psi-3)}+1]}+$$

$$----------$$

$$T_{[(n(n+1))^4]} + T_{[(n(n+1))^4+1]}+$$

$$T_{[(n(n+1))^2]} + T_{[(n(n+1))^2+1]} + T_{[n(n+1)^2-1]} + T_{n(n+1)} + T_{\frac{[n(n+1)]}{2}} + T_n$$

$$= T_{[(n(n+1))^{2(\psi-1)}+\cdots+(n(n+1))^4+(n(n+1))^2+\frac{3n(n+1)}{2}]} \cdot \quad [n \geq 1]$$

proving the theorems.

This Proves the General theorem

Theorem 6.9. *For any Natural Number $N \geq 2$, there Exist Infinite Triangular Numbers that are the Sum of N Triangular Numbers each.*

 Proof. For $N = 2$ we have the standard result. (Ref. 6)

For $N > 2$, we have proved the result.

For $N = 3, 5, 7, \cdots$ and $N = 4, 6, 8, \cdots$ in Theorems 6) and Theorems 6a) respectively. This Completes the Proof. $\square$

7. Theorems on Two Special Classes of Triangular Numbers

Theorem 7.1. *If* $x = \dfrac{n(n+2\alpha+1)}{2} + 1$, $y = n+1$, $z = \dfrac{n(n+2\alpha+1)}{2}$

$$T_x = T_y + T_z + (\alpha-1)n = T_y + T_z + (\alpha-1)(y-1)$$

$$[\alpha = 0, 1, 2, \cdots, \alpha] \qquad [n = 0, 1, 2, \cdots, \alpha]$$

Proof. If $x = \dfrac{n(n+2\alpha+1)}{2} + 1$, $y = n+1$, $z = \dfrac{n(n+2\alpha+1)}{2}$

7–1–1)   For, $\alpha = 0$ we can prove by inducting on "n"

$$x = \frac{n(n+1)}{2} + 1, \quad y = n+1, \quad z = \frac{n(n+1)}{2}$$

$$T_x = T_y + T_z - n = T_y + T_z - (y-1) \quad [n = 0, 1, 2, \cdots, \infty]$$

7–1–2) For, $\alpha = 1$ we can prove by inducting on "n"

$$x = \frac{n(n+3)}{2} + 1, \quad y = n+1, \quad z = \frac{n(n+3)}{2}$$
$$T_x = T_y + T_z \quad [n = 0, 1, 2, \cdots, \infty]$$

which is the standard result that states that there exist infinite Triangular Numbers that are the sum of two other Triangular Numbers. [Ref. 6]

7–1–3) For, $\alpha = 2$ we can prove by inducting on "n"

$$x = \frac{n(n+5)}{2} + 1, \quad y = n+1, \quad z = \frac{n(n+5)}{2}$$
$$T_x = T_y + T_z + n = T_y + T_z + (y-1) \quad [n = 0, 1, 2, \cdots, \infty]$$

7–1–4) For, $\alpha = 3$ we can prove by inducting on "n"

$$x = \frac{n(n+7)}{2} + 1, \quad y = n+1, \quad z = \frac{n(n+7)}{2}$$
$$T_x = T_y + T_z + n = T_y + T_z + (y-1) \quad [n = 0, 1, 2, \cdots, \infty]$$

7–1–5) For, $\alpha = 4$ we can prove by inducting on "n"

$$x = \frac{n(n+9)}{2} + 1, \quad y = n+1, \quad z = \frac{n(n+9)}{2}$$
$$T_x = T_y + T_z + n = T_y + T_z + (y-1) \quad [n = 0, 1, 2, \cdots, \infty]$$

7–1–6) For, $\alpha = 5$ we can prove by inducting on "n"

$$x = \frac{n(n+11)}{2} + 1, \quad y = n+1, \quad z = \frac{n(n+11)}{2}$$
$$T_x = T_y + T_z + n = T_y + T_z + (y-1) \quad [n = 0, 1, 2, \cdots, \infty]$$

7–1–k+1) For, $\alpha = k$ we can prove by inducting on "n"

$$x = \frac{n(n+2k+1)}{2} + 1, \quad y = n+1, z = \frac{n(n+2k+1)}{2}$$
$$T_x = T_y + T_z + (k-1)n = T_y + T_z + (k-1)(y-1)$$
$$[k = 0, 1, 2, \cdots, \infty] \, [n = 0, 1, 2, \cdots, \infty]$$

7–1–k+2) For, $\alpha = k+1$ we can prove by inducting on "n"

$$x = \frac{n(n+2(k+1)+1)}{2} + 1, \quad y = n+1, z = \frac{n(n+2(k+1)+1)}{2}$$
$$T_x = T_y + T_z + (k)n = T_y + T_z + (k)(y-1)$$
$$[k = 0, 1, 2, \cdots, \infty] \, [n = 0, 1, 2, \cdots, \infty]$$

thus Completing the Induction on "α" proving the theorem that, if

$$x = \frac{n(n + 2\alpha + 1)}{2} + 1, \quad y = n + 1, \ z = \frac{n(n + 2\alpha + 1)}{2}$$
$$T_x = T_y + T_z + (\alpha - 1)n = T_y + T_z + (\alpha - 1)(y - 1)$$
$$[\alpha = 0, 1, 2, \cdots, \infty]$$

$[n = 0, 1, 2, \cdots, \infty]$ $\square$

Theorem 7.2. *If*

$$x = \frac{(n + 1)(n + 2\beta)}{2} + 1, \quad y = n + 1, z = \frac{(n + 1)(n + 2\beta)}{2}$$
$$T_x = T_y + T_z + \beta + (\beta - 1)n = T_y + T_z + \beta + (\beta - 1)(y - 1)$$
$$[\beta = 1, 2, 3, \cdots, \infty] \ [n = 0, 1, 2, \cdots, \infty]$$

Proof. If

$$x = \frac{(n + 1)(n + 2\beta)}{2} + 1, \quad y = n + 1, z = \frac{(n + 1)(n + 2\beta)}{2}.$$

7–2–1) For, $\beta = 1$ we can prove by inducting on "n"

$$x = \frac{(n + 1)(n + 2)}{2} + 1, \quad y = n + 1, z = \frac{(n + 1)(n + 2)}{2}$$
$$T_x = T_y + T_z + 1 \quad [n = 0, 1, 2, \cdots, \infty]$$

7–2–2) For, $\beta = 2$ we can prove by inducting on "n"

$$x = \frac{(n + 1)(n + 4)}{2} + 1, \quad y = n + 1, z = \frac{(n + 1)(n + 4)}{2}$$
$$T_x = T_y + T_z + 2 + n = T_y + T_z + 2 + (y - 1) \quad [n = 0, 1, 2, \cdots, \infty]$$

7–2–3) For, $\beta = 3$ we can prove by inducting on "n"

$$x = \frac{(n + 1)(n + 6)}{2} + 1, \quad y = n + 1, z = \frac{(n + 1)(n + 6)}{2}$$
$$T_x = T_y + T_z + 3 + 2n = T_y + T_z + 3 + 2(y - 1) \quad [n = 0, 1, 2, \cdots, \infty]$$

7–2–4) For, $\beta = 4$ we can prove by inducting on "n"

$$x = \frac{(n + 1)(n + 8)}{2} + 1, \quad y = n + 1, z = \frac{(n + 1)(n + 8)}{2}$$
$$T_x = T_y + T_z + 4 + 3n = T_y + T_z + 4 + 3(y - 1) \ [n = 0, 1, 2, \cdots, \infty]$$

7–2–5) For, $\beta = 5$ we can prove by inducting on "n"

$$x = \frac{(n + 1)(n + 10)}{2} + 1, \quad y = n + 1, z = \frac{(n + 1)(n + 10)}{2}$$
$$T_x = T_y + T_z + 5 + 4n = T_y + T_z + 5 + 4(y - 1) \quad [n = 0, 1, 2, \cdots, \infty]$$

− − − − − − −

7–2–k) For, $\beta = k$ we can prove by inducting on "n"

$$x = \frac{(n+1)(n+2k)}{2} + 1, \quad y = n+1, z = \frac{(n+1)(n+2k)}{2}$$

$$T_x = T_y + T_z + k + (k-1)n = T_y + T_z + k + (k-1)(y-1)$$

$$[k = 1, 2, 3, \cdots, \infty] \; [n = 0, 1, 2, \cdots, \infty]$$

7–2–k) For, $\beta = k + 1$ we can prove by inducting on "n"

$$x = \frac{(n+1)(n+2(k+1))}{2} + 1, \quad y = n+1, z = \frac{(n+1)(n+2(k+1))}{2}$$

$$T_x = T_y + T_z + (k+1) + (k)n = T_y + T_z + (k+1) + (k)(y-1)$$

$$[k = 1, 2, 3, \cdots, \infty] \; [n = 0, 1, 2, \cdots, \infty]$$

thus completing the induction on "β" proving the theorem that

If

$$x = \frac{(n+1)(n+2\beta)}{2} + 1, \quad y = n+1, z = \frac{(n+1)(n+2\beta)}{2}$$

$$T_x = T_y + T_z + \beta + (\beta-1)n = T_y + T_z + \beta + (\beta-1)(y-1)$$

$$[\beta = 1, 2, 3 \cdots, \infty] \; [n = 0, 1, 2, \cdots, \infty]$$

$\square$

We come to the end of this paper. The Super-Sums of each Sub-Class of Triangular Numbers upto any Level and other interesting things are possible. [Ref. 7,8,9.]

BIBLIOGRAPHY

1) Pythagoras.   (568–500 B. C)   [□]
2) Nicomachus.   (circa. 100 A. D)   [□]
3) Aryabhatta.   (circa. 500 A. D)   [□]
4) Pascal, Blaise.   (1623–1662)   [□]
5) Euler, Leonard.   (1707 –1783)   [□]
   [□] = from 6)
6) David M. Burton:
   ELEMENTARY NUMBER THEORY.
   UNIVERSAL BOOK STALL. NEW DELHI. Second Edition. Reprint 1998.
   (p. 18–19)
7) Narayanan Raghunathan
   Generalized arithmetic progressions – a survey [Submitted]
8) Narayanan Raghunathan
   Generalized geometric progressions – a survey [Submitted]

9) Narayanan Raghunathan.
    THE UNIVERSES OF TRIANGULAR NUMBER RHYTHMS. [*Getting Ready*]

Infinite Sequential Averages

(Dedicated to my brother Anand)

Abstract. Formula for the nth term of the Sequential Averages

$$[{}_{[A]}^{S}]n = \frac{1 \cdot 2 \cdot 3 \cdot \ldots, \cdot A}{A}, \frac{2 \cdot 3 \cdot 4 \cdot \ldots, \cdot (A+1)}{A},$$

$$\frac{3 \cdot 4 \cdot 5 \cdot \ldots, \cdot (A+2)}{A}, \cdots,$$

$$\frac{n(n+1)(n+2)\ldots \cdot (n+A-1)}{A}$$

is determined. Formula for the Sum up to n terms of the Sequential Averages is also proved. For $A = 2$ we have the traditional result for Triangular Numbers. Infinite Types of unique sub-classes of Sequential Averages are defined and the formulae for their n^{th} terms are derived and these are expressed in terms of binomial coefficients also.

Key Words:- Infinite Sequences, Mathematical Induction.
AMS Subject Classification No:- 11A99.

1. The Sequential Averages and the nth Terms

$$[{}_{[A]}^{S}]n = \frac{1 \cdot 2 \cdot 3 \cdot \ldots, \cdot A}{A}, \frac{2 \cdot 3 \cdot 4 \cdot \ldots, \cdot (A+1)}{A},$$

$$\frac{3 \cdot 4 \cdot 5 \cdot \ldots, \cdot (A+2)}{A}, \cdots,$$

$$\frac{n(n+1)(n+2)\ldots \cdot (n+A-1)}{A}$$

is the given sequence of Sequential Averages. $A =$ Average Level of the given Sequential Average.

When $A = 2$, we have the case of Triangular Numbers.

Theorem 1.1.

$$[{}_{[A]}^{S}]n = [{}_{[A]}^{S}]n^{[1]} \, a_1 + a_2(n-1) + a_3(n-2) + \cdots + a_A(n-(A-1))$$

$$+ a_{A+1}\left[\frac{(n-A+1)(n-A)}{2}\right]$$

where

$$a_1 = (A-1)!, a_2 = A!, \ a_3 = \frac{A!(A-1)}{2}, \ a_4 = \frac{A!(A-2)}{2},$$

$$a_5 = \frac{A!(A-3)}{2}, \ \text{------} \ a_A = \frac{A!(A-(A-2))}{2} = A!,$$

$$a_{A+1} = (A-1)!. \tag{1.1}$$

$\{In \ [\underset{[A]}{S}]n = [\underset{[A]}{\overset{[1]}{S}}]n, \ "A" \ stands \ for \ the \ Average\text{-}Level \ Of \ the \ given \ Se$-
quential Average and the superscript (though not essential here) indicates the
Super-Sum Level Sequential-Averages. [See. 2]

Proof. The Proof follows by the Traditional Method of Mathematical Induction
on "n" for each A-Level and on "A" itself for the General Result.
For $A = 2$ we have

$$[\underset{[2]}{S}]n = a_1 + a_2(n-1) + a_3\left[\frac{(n-1)(n-2)}{2}\right] = [\underset{[2]}{\overset{[1]}{S}}]n$$

$$a_1 = 1! = 1, a_2 = 2! = 2, a_3 = 2! \cdot \frac{1}{2} = 1$$

$$[\underset{[2]}{S}]n = 1 \cdot 2, 2 \cdot 3, 3 \cdot 4, \cdots, n(n+1) = [\underset{[2]}{\overset{[1]}{S}}]n$$

$$[\underset{[2]}{S}]n = 1! + 2!(n-1) + \frac{2! \cdot 1!}{2}\left[\frac{(n-1)(n-2)}{2}\right]$$

$$= 1 + 2(n-1) + \left[\frac{(n-1)(n-2)}{2}\right] = [\underset{[2]}{\overset{[1]}{S}}]n$$

$$[\underset{[A]}{S}]n = 1 \cdot 2 \cdot 3 \cdot \ \text{---} \ \cdot A, \ 2 \cdot 3 \cdot 4 \cdot \ \text{---} \ \cdot (A+1),$$

$$3 \cdot 4 \cdot 5 \cdot \ \text{---} \ \cdot (A+2), \ \text{---},$$

$$n(n+1)(n+2) \ \text{---} \ \cdot (n+A-1).[\underset{[A]}{\overset{[1]}{S}}]n$$

$$[\underset{[A]}{S}]n = a_1 + a_2(n-1) + a_3(n-2) + \ \text{---} + a_A(n-(A-1))$$

$$+ a_{A+1}\left[\frac{(n-A+1)(n-A)}{2}\right] = [\underset{[A]}{\overset{[1]}{S}}]n \qquad [A = 2, 3, 4 \text{---} \infty]$$

$$(A-1)! + A!(n-1) + \frac{A!(A-1)(n-2)}{2} + \ \text{---}$$

$$+ \, A!\frac{(A-(A-2))(n-(A-1))+(A-1)!}{2}$$

$$\left[\frac{(n-A+1)(n-A)}{2}\right] = [\,{\overset{[1]}{\underset{[A]}{S}}}\,]n$$

$$[\,{\overset{[1]}{\underset{[A]}{S}}}\,]n = n(n+1)(n+2)\cdots\cdot(n+A-1)$$

$$= (A-1)! + A!(n-1) + A!\frac{(A-1)(n-2)}{2} + \cdots$$

$$+ \, A!\frac{(A-(A-2))(n-(A-1))+(A-1)!}{2}$$

$$[\frac{(n-A+1)(n-A)}{2}] = [\,{\overset{[1]}{\underset{[A]}{S}}}\,]n. \qquad\qquad \square$$

2. Formula for the Sum up to n Terms of the Sequential

Averages:

Let S_n be the Sum upto n terms of the Sequential Averages

$$S_n\left[\,[\,{\overset{[1]}{\underset{[2]}{S}}}\,]\,\right] = \frac{1\cdot 2}{2} + \frac{2\cdot 3}{2} + \frac{3\cdot 4}{2} + \cdots + \frac{n(n+1)}{2} =$$

$$\sum_{k=1}^{n} k(k+1) = \frac{n(n+1)(n+2)}{2\cdot 3} = 1!\left[\frac{n(n+1)(n+2)}{3!}\right] = [\,{\overset{[2]}{\underset{[2]}{S}}}\,]n$$

This is a well known identity. [Ref: 1]

The Following Identities can be Inductively Derived !

$$S_n\left[\,[\,{\overset{[1]}{\underset{[3]}{S}}}\,]\,\right] = \frac{1\cdot 2\cdot 3}{3} + \frac{2\cdot 3\cdot 4}{3} + \frac{3\cdot 4\cdot 5}{3} + \cdots + \frac{n(n+1)(n+2)}{3}$$

$$= \sum_{k=1}^{n} \frac{k(k+1)(k+2)}{3} = \frac{n(n+1)(n+2)(n+3)}{3\cdot 4}$$

$$= 2!\left[\frac{n(n+1)(n+2)(n+3)}{4!}\right] = [\,{\overset{[2]}{\underset{[3]}{S}}}\,]n$$

$$S_n\left[\,[\,{\overset{[1]}{\underset{[A]}{S}}}\,]\,\right] = \frac{1\cdot 2\cdot 3\cdot \,\cdots\,\cdot A}{A} + \frac{2\cdot 3\cdot 4\cdot \,\cdots\,\cdot (A+1)}{A}$$

$$+ \frac{3\cdot 4\cdot 5\cdot \,\cdots\,\cdot (A+2)}{A} + \cdots$$

$$+ \frac{n(n+1)(n+2)\,\cdots\,\cdot (n+A-1)}{A}$$

$$= \sum_{k=1}^{n} \frac{k(k+1)(k+2)\cdots(k+A-1)}{A}$$

$$= \frac{n(n+1)(n+2)\cdots(n+A)}{A(A+1)}$$

$$= (A-1)! \left[\frac{n(n+1)(n+2)\cdots(n+A)}{(A+1)!} \right] = [\,\substack{[2]\\S\\[A]}\,]n. \qquad (2.1)$$

We can of course determine a Formula for the Super-Sum-α-Level by proceeding on inducting on α. Here $\alpha = 1$ and we have determined the Second-Level of the Given Sequential Averages See. Ref: 2,4]

3. Another Equality

From (1.1) we have

$$[\,\substack{S\\[A]}\,]n = [\,\substack{[1]\\S\\[A]}\,]n = a_1 + a_2(n-1) + a_3(n-2) + \cdots$$

$$+ a_A(n-(A-1)) + a_{A+1}\left[\frac{(n-A+1)(n-A)}{2}\right]$$

$$a_1 = (A-1)!, a_2 = A!, \ a_3 = A!\frac{(A-1)}{2}, \ a_4 = A!\frac{(A-2)}{2},$$

$$a_5 = A!\frac{(A-3)}{2}\cdots\cdots a_A = A!\frac{(A-(A-2))}{2}$$

$$= A!, a_{A+1} = (A-1)!.$$

Now we can easily prove by Induction that the sum of this progression up to n terms is

$$S_n \begin{bmatrix}[1]\\S\\[A]\end{bmatrix} = a_1 n + a_2\left[\frac{n(n-1)}{2}\right] + a_3\left[\frac{(n-1)(n-2)}{2}\right] + \cdots$$

$$+ a_A\left[\frac{(n-A+1)(n-A)}{2}\right] + a_{A+1}\sum_{1}^{n-A} i\frac{(n-i)(n-(i+1))}{2}$$

$$= a_1 n + a_2\left[\frac{n(n-1)}{2!}\right] + a_3\left[\frac{(n-1)(n-2)}{2!}\right] + \cdots$$

$$+ a_A\left[\frac{(n-A+1)(n-A)}{2!}\right]$$

$$+ a_{A+1}\left[\frac{(n-A+2)(n-A+1)(n-A)}{3!}\right]$$

$$= (A-1)![n] + A! \left[\frac{n(n-1)}{2!} \right] + A! \frac{(A-1)}{2} \left[\frac{(n-1)(n-2)}{2!} \right] + \cdots$$

$$+ A! \left[\frac{(n-A+1)(n-A)}{2!} \right]$$

$$+ (A-1)! \left[\frac{(n-A+2)(n-A+1)(n-A)}{3!} \right]$$

$$= (A-1)! \left[\frac{n(n+1)(n+2) \cdots (n+A)}{(A+1)!} \right] \text{ [Substituting (2.1)]}$$

Dividing both sides by $(A-1)!$ we have

$$= \frac{1}{(A-1)!} \left[(A-1)![n] + A! \left[\frac{n(n-1)}{2!} \right] \right.$$

$$+ A! \frac{(A-1)}{2} \left[\frac{(n-1)(n-2)}{2!} \right] + \cdots + A! \left[\frac{(n-A+1)(n-A)}{2!} \right]$$

$$\left. + (A-1)! \left[\frac{(n-A+2)(n-A+1)(n-A)}{3!} \right] \right]$$

$$= \left[\frac{n(n+1)(n+2) \cdots (n+A)}{(A+1)!} \right] = \left[\begin{matrix} [2] \\ S \\ [A] \end{matrix} \right]_n$$

We can of course determine a Formula for the Super-Sum-α-Level by proceeding on inducting on α. Here $\alpha = 1$ and we have determined the Second-Level of the Given Sequential Averages See. Ref:2,4]

4. Infinite Root Levels and Layer Levels of Sequential

Averages.

In this section we define various Root-Levels and Layer-Levels of Sequential Averages and investigate them.

In $^L S_n^{[R]}{}_{[A]}$, R may be called the Root-Level and L the Layer level which define a specific sub-class of Sequential Averages.

The sequence of Layer-Levels for each Root-Level Facilitates the Proof for All Root-Levels.

We can easily prove by Repeated Induction the following results.

$$^1 S_{n+1}^{[1]} = \frac{(n+1)(n+2)}{2} \quad [\, n = 0, 1, 2 \cdots \infty \,]$$

[2]

$$= S_1, S_2, S_3, S_4, S_5, ---, S_{1+(n-1)}, ---\infty$$
$$\quad [2]\ [2]\ [2]\ [2]\ [2] \qquad\quad [2]$$

$$^{[1]}_{2}S_{n+1}_{[2]} = \frac{(2n+1)(2n+2)}{2} \qquad [\,n = 0, 1, 2 ---\infty\,]$$

$$= S_1, S_3, S_5, S_7, S_9, ---, S_{1+2(n-1)}, ---\infty$$
$$\quad [2]\ [2]\ [2]\ [2]\ [2] \qquad\quad [2]$$

$$^{[1]}_{\phi}S_{n+1}_{[2]} = \frac{(\phi n+1)(\phi n+2)}{2} \qquad \begin{array}{l}[\,n = 0, 1, 2 ---\infty\,]\\ [\,\phi = 1, 2, 3 ---\infty\,]\end{array}$$

$$= S_1, S_{1+\phi}, S_{1+2\phi}, S_{1+3\phi}, S_{1+4\phi}, ---,$$
$$\quad [2]\ [2]\qquad [2]\qquad [2]\qquad [2]$$

$$S_{1+\phi(n-1)}, ---\infty$$
$$[2]$$

$$^{[2]}_{1}S_{n+1}_{[2]} = \frac{(2n+1)(2n+2)}{2} \qquad [\,n = 0, 1, 2 ---\infty\,]$$

$$= S_1, S_3, S_5, S_7, S_9, ---, S_{1+2(n-1)}, ---\infty$$
$$\quad [2]\ [2]\ [2]\ [2]\ [2] \qquad\quad [2]$$

$$^{[2]}_{\phi}S_{n+1}_{[2]} = \frac{(2\phi n+1)(2\phi n+2)}{2} \qquad \begin{array}{l}[\,n = 0, 1, 2 ---\infty\,]\\ [\,\phi = 1, 2, 3 ---\infty\,]\end{array}$$

$$= S_1, S_{1+2\phi}, S_{1+4\phi}, S_{1+6\phi}, S_{1+8\phi}, ---,$$
$$\quad [2]\ [2]\qquad [2]\qquad [2]\qquad [2]$$

$$S_{1+2\phi(n-1)}, ---\infty$$
$$[2]$$

$$^{[R]}_{1}S_{n+1}_{[2]} = \frac{(Rn+1)(Rn+2)}{2} \qquad \begin{array}{l}[\,n = 0, 1, 2 ---\infty\,]\\ [\,R = 1, 2, 3 ---\infty\,]\end{array}$$

$$= S_1, S_{1+R}, S_{1+2R}, S_{1+3R}, S_{1+4R}, ---,$$
$$\quad [2]\ [2]\qquad [2]\qquad [2]\qquad [2]$$

$$S_{1+R(n-1)}, ---\infty$$
$$[2]$$

$$\overset{[R]}{\underset{[2]}{2S_{n+1}}} = \frac{(2Rn+1)(2Rn+2)}{2} \qquad \begin{array}{l} [\ n=0,1,2\text{---}\infty\ ] \\ [\ R=1,2,3\text{---}\infty\ ] \end{array}$$

$$= \underset{[2]}{S_1}\ ,\ \underset{[2]}{S_{1+2R}},\ \underset{[2]}{S_{1+4R}},\ \underset{[2]}{S_{1+6R}},\ \underset{[2]}{S_{1+8R}},\ \text{---},$$

$$\underset{[2]}{S_{1+2R(n-1)}},\ \text{---}\infty$$

- - - - - - - - - - - - - - -

$$\overset{[R]}{\underset{[2]}{\phi S_{n+1}}} = \frac{(\phi Rn+1)(\phi Rn+2)}{2} \qquad \begin{array}{l} [\ n=0,1,2\text{---}\infty\ ] \\ [\ \phi=1,2,3\text{---}\infty\ ] \\ [\ R=1,2,3\text{---}\infty\ ] \end{array}$$

$$= \underset{[2]}{S_1},\ \underset{[2]}{S_{1+\phi R}},\ \underset{[2]}{S_{1+2\phi R}},\ \underset{[2]}{S_{1+3\phi R}},\ \underset{[2]}{S_{1+4\phi R}},\ \text{---},$$

$$\underset{[2]}{S_{1+\phi R(n-1)}},\ \text{---}\infty \qquad \text{See Ref}:3]$$

- - - - - - - - - - - - - - -

$$\overset{[3]}{\underset{[3]}{\phi S_{n+1}}} = \frac{(3\phi n+1)(3\phi n+2)(3\phi n+3)}{3} \quad \begin{array}{l} [\ n=0,1,2\text{---}\infty\ ] \\ [\ \phi=1,2,3\text{---}\infty\ ] \end{array}$$

$$= \underset{[3]}{S_1},\ \underset{[3]}{S_{1+3\phi}},\ \underset{[3]}{S_{1+6\phi}},\ \underset{[3]}{S_{1+9\phi}},\ \underset{[3]}{S_{1+12\phi}},\ \text{---},$$

$$\underset{[3]}{S_{1+3\phi(n-1)}},\ \text{---}\infty$$

- - - - - - - - - - - - - - -

$$\overset{[R]}{\underset{[3]}{1S_{n+1}}} = \frac{(Rn+1)(Rn+2)(Rn+3)}{3} \qquad \begin{array}{l} [\ n=0,1,2\text{---}\infty\ ] \\ [\ R=1,2,3\text{---}\infty\ ] \end{array}$$

$$= \underset{[3]}{S_1}\ ,\ \underset{[3]}{S_{1+R}},\ \underset{[3]}{S_{1+2R}},\ \underset{[3]}{S_{1+3R}},\ \underset{[3]}{S_{1+4R}},\ \text{---},$$

$$\underset{[3]}{S_{1+R(n-1)}},\ \text{---}\infty$$

$$\overset{[R]}{\underset{[3]}{2S_{n+1}}} = \frac{(2Rn+1)(2Rn+2)(2Rn+3)}{3} \quad \begin{array}{l} [\ n=0,1,2\text{---}\infty\ ] \\ [\ R=1,2,3\text{---}\infty\ ] \end{array}$$

$$= \underset{[3]}{S_1}\ ,\ \underset{[3]}{S_{1+2R}},\ \underset{[3]}{S_{1+4R}},\ \underset{[3]}{S_{1+6R}},\ \underset{[3]}{S_{1+8R}},\ \text{---},$$

$$\underset{[3]}{S_{1+2R(n-1)}},\ \text{---}\infty$$

- - - - - - - - - - - - - - -

$$\begin{array}{l}[R]\\ {}^{\phi}S_{n+1} \\ {}_{[3]}\end{array} = \frac{(\phi Rn+1)(\phi Rn+2)(\phi Rn+3)}{3} \begin{array}{l}[\,n=0,1,2\,\text{---}\infty\,]\\ [\,\phi=1,2,3\,\text{---}\infty\,]\\ [\,R=1,2,3\,\text{---}\infty\,]\end{array}$$

$$= \underset{[3]}{S_1}, \underset{[3]}{S_{1+\phi R}}, \underset{[3]}{S_{1+2\phi R}}, \underset{[3]}{S_{1+3\phi R}}, \underset{[3]}{S_{1+4\phi R}}, \text{---},$$

$$\underset{[3]}{S_{1+\phi R(n-1)}}, \text{---}\infty$$

$$\begin{array}{l}[1]\\ {}^{1}S_{n+1} \\ {}_{[A]}\end{array} = \frac{(n+1)(n+2)\text{---}(n+A)}{A} \qquad \begin{array}{l}[\,n=0,1,2\,\text{---}\infty\,]\\ [\,A=2,3,4\,\text{---}\infty]\end{array}$$

$$= \underset{[A]}{S_1}, \underset{[A]}{S_2}, \underset{[A]}{S_3}, \underset{[A]}{S_4}, \underset{[A]}{S_5}, \text{---}, \underset{[A]}{S_{1+(n-1)}}, \text{---}\infty$$

$$\begin{array}{l}[1]\\ {}^{2}S_{n+1} \\ {}_{[A]}\end{array} = \frac{(2n+1)(2n+2)\text{---}(2n+A)}{A} \begin{array}{l}[\,n=0,1,2\,\text{---}\infty\,]\\ [\,A=2,3,4\,\text{---}\infty]\end{array}$$

$$= \underset{[A]}{S_1}, \underset{[A]}{S_3}, \underset{[A]}{S_5}, \underset{[A]}{S_7}, \underset{[A]}{S_9}, \text{---}, \underset{[A]}{S_{1+2(n-1)}}, \text{---}\infty$$

$$\begin{array}{l}[1]\\ {}^{\phi}S_{n+1} \\ {}_{[A]}\end{array} = \frac{(\phi n+1)(\phi n+2)\text{---}(\phi n+A)}{A} \begin{array}{l}[\,n=0,1,2\,\text{---}\infty\,]\\ [\,\phi=1,2,3\,\text{---}\infty\,]\\ [\,A=2,3,4\,\text{---}\infty]\end{array}$$

$$= \underset{[A]}{S_1}, \underset{[A]}{S_{1+\phi}}, \underset{[A]}{S_{1+2\phi}}, \underset{[A]}{S_{1+3\phi}}, \underset{[A]}{S_{1+4\phi}}, \text{---},$$

$$\underset{[A]}{S_{1+\phi(n-1)}}, \text{---}\infty$$

$$\begin{array}{l}[2]\\ {}^{\phi}S_{n+1} \\ {}_{[A]}\end{array} = \frac{(2\phi n+1)(2\phi n+2)\text{---}(2\phi n+A)}{A}$$

$$\begin{array}{l}[n=0,1,2\,\text{---}\infty]\\ [\phi=1,2,3\,\text{---}\infty]\\ [A=2,3,4\,\text{---}\infty]\end{array}$$

$$= \underset{[A]}{S_1}, \underset{[A]}{S_{1+2\phi}}, \underset{[A]}{S_{1+4\phi}}, \underset{[A]}{S_{1+6\phi}}, \underset{[A]}{S_{1+8\phi}}, \text{---},$$

$$\underset{[A]}{S_{1+2\phi(n-1)}}, \text{---}\infty$$

$$\begin{array}{l} {}^{[R]}_1 S_{n+1} \\ {}_{[A]} \end{array} = \frac{(Rn+1)(Rn+2)\text{---}(Rn+A)}{A}$$

$$[n = 0, 1, 2\text{---}\infty]$$
$$[R = 1, 2, 3\text{---}\infty]$$
$$[A = 2, 3, 4\text{---}\infty]$$

$$= \underset{[A]}{S_1}, \underset{[A]}{S_{1+R}}, \underset{[A]}{S_{1+2R}}, \underset{[A]}{S_{1+3R}}, \underset{[A]}{S_{1+4R}}, \text{---},$$

$$\underset{[A]}{S_{1+R(n-1)}}, \text{---}\infty$$

$$\begin{array}{l} {}^{[R]}_2 S_{n+1} \\ {}_{[A]} \end{array} = \frac{(2Rn+1)(2Rn+2)\text{---}(2Rn+A)}{A}$$

$$[n = 0, 1, 2\text{---}\infty]$$
$$[R = 1, 2, 3\text{---}\infty]$$
$$[A = 2, 3, 4\text{---}\infty]$$

$$= \underset{[A]}{S_1}, \underset{[A]}{S_{1+2R}}, \underset{[A]}{S_{1+4R}}, \underset{[A]}{S_{1+6R}}, \underset{[A]}{S_{1+8R}}, \text{---},$$

$$\underset{[A]}{S_{1+2R(n-1)}}, \text{---}\infty$$

- - - - - - - - - - - - - - -

$$\begin{array}{l} {}^{[R]}_{\phi} S_{n+1} \end{array} = \frac{(\phi Rn+1)(\phi Rn+2)\text{---}(\phi Rn+A)}{A}$$

$$[n = 0, 1, 2\text{---}\infty]$$
$$[\phi = 1, 2, 3\text{---}\infty]$$
$$[R = 2, 3, 4\text{---}\infty]$$
$$[A = 1, 2, 3\text{---}\infty]$$

$$= \underset{[A]}{S_1}, \underset{[A]}{S_{1+\phi R}}, \underset{[A]}{S_{1+2\phi R}}, \underset{[A]}{S_{1+3\phi R}}, \underset{[A]}{S_{1+4\phi R}}, \text{---},$$

$$\underset{[A]}{S_{1+\phi R(n-1)}}, \text{---}\infty$$

5. The Infinite Sequential-Average Classes in Terms of Binomial Coefficients

We can easily prove by Repeated Induction the following results.

$$\begin{array}{l} {}^{[1]}_1 S_{n+1} \\ {}_{[2]} \end{array} = \binom{n+2}{2} = T_{n+1} = \text{Triangular Numbers } [n \geq 0]$$

[This is the standard result. Ref. 1)]

$$\begin{array}{c}[1]\\ {}^{2}S_{n+1}\\ [2]\end{array} = \binom{2n+2}{2}$$

$$\begin{array}{c}[1]\\ {}^{\phi}S_{n+1}\\ [2]\end{array} = \binom{\phi n+2}{2}$$

$$\begin{array}{c}[R]\\ {}^{1}S_{n+1}\\ [2]\end{array} = \binom{Rn+2}{2}$$

$$\begin{array}{c}[R]\\ {}^{2}S_{n+1}\\ [2]\end{array} = \binom{2Rn+2}{2}$$

$$\begin{array}{c}[R]\\ {}^{\phi}S_{n+1}\\ [2]\end{array} = \binom{\phi Rn+2}{2}$$

$$\begin{array}{c}[1]\\ {}^{1}S_{n+1}\\ [3]\end{array} = (2!)\binom{n+3}{3}$$

$$\begin{array}{c}[1]\\ {}^{2}S_{n+1}\\ [3]\end{array} = (2!)\binom{2n+3}{3}$$

$$\begin{array}{c}[1]\\ {}^{\phi}S_{n+1}\\ [3]\end{array} = (2!)\binom{\phi n+3}{3}$$

$$\begin{array}{c}[R]\\ {}^{1}S_{n+1}\\ [3]\end{array} = (2!)\binom{Rn+3}{3}$$

$$\begin{array}{c}[R]\\ {}^{\phi}S_{n+1}\\ [3]\end{array} = (2!)\binom{\phi Rn+3}{3}$$

$$\begin{matrix} [1] \\ {}^{1}S_{n+1} \\ [A] \end{matrix} = ((A-1)!)\begin{pmatrix} n+A \\ A \end{pmatrix}$$

$$\begin{matrix} [1] \\ {}^{2}S_{n+1} \\ [A] \end{matrix} = ((A-1)!)\begin{pmatrix} 2n+A \\ A \end{pmatrix}$$

- - - - - - - - - - -

$$\begin{matrix} [1] \\ {}^{\phi}S_{n+1} \\ [A] \end{matrix} = ((A-1)!)\begin{pmatrix} \phi n+A \\ A \end{pmatrix}$$

- - - - - - - - - - -

$$\begin{matrix} [R] \\ {}^{1}S_{n+1} \\ [A] \end{matrix} = ((A-1)!)\begin{pmatrix} Rn+A \\ A \end{pmatrix}$$

- - - - - - - - - - -

$$\begin{matrix} [R] \\ {}^{\phi}S_{n+1} \\ [A] \end{matrix} = ((A-1)!)\begin{pmatrix} \phi Rn+A \\ A \end{pmatrix}$$

$$[\, n = 0,1,2\,\text{-}\text{-}\text{-}\infty \,]$$
$$[\, \phi = 1,2,3\,\text{-}\text{-}\text{-}\infty \,]$$
$$[\, R = 2,3,4\,\text{-}\text{-}\text{-}\infty \,]$$
$$[\, A = 1,2,3\,\text{-}\text{-}\text{-}\infty]$$

We can of course find formulae for the Sums and Super-Sums of Sequential Averages upto any α-Level notated as below.

$$S_{n+1}\begin{bmatrix} [\alpha] \\ [R] \\ {}^{\phi}S \\ [A] \end{bmatrix}$$

$\phi = $ Layer Level
$R = $ Root Level
$A = $ Average Level
$\alpha = $ Super-Sum Level
$S_{n+1} = $ Sum upto $n+1$ terms $[\, n \geq 0 \,]$

[For more details See Ref :2,3,4,5.]

Bibliography

1] David M. Burton:
ELEMENTARY NUMBER THEORY.
UNIVERSAL BOOK STALL. NEW DELHI. Second Edition. Reprint 1998. (Ch. 1)
2] Narayanan Raghunathan :
Generalized Arithmetic Progressions - A Survey. [Submitted]

3] Narayanan Raghunathan :
   Triangular numbers - Some General Theorems and Related Results. [Submitted]

4] Narayanan Raghunathan :
   Functions and their Progressions - An Elementary Text. [*unpublished*]

5] Narayanan Raghunathan :
   Infinite Universes of Sequential Averages. [*being compiled*]

A Generalized Elementary Sequence and Its Sum

(Dedicated to my brother Anand)

Abstract. Formula for the nth term of the Sequence

$$[\textstyle\coprod]_n \atop [\lambda] = 1 \cdot 2 \cdot 3 \text{ --- } \lambda, \; 2 \cdot 3 \cdot 4 \cdot \text{ ---- } \cdot (\lambda+1), 3 \cdot 4 \cdot 5 \cdot \text{ --- } \cdot (\lambda+2),$$

$$\text{ --- }, \; n(n+1)(n+2) \text{ --- } \cdot (n+\lambda-1) \cdot$$

is determined. Formula for the Sum up to n terms of the Sequence is also proved. For $\lambda = 2$ we have the traditional result.

Key Words:- Infinite Sequences, Mathematical Induction.
AMS Subject Classification No:- 11A99.

1. The Sequence and the nth Term

$$[\textstyle\coprod]_n \atop [\lambda] = 1 \cdot 2 \cdot 3 \cdot \text{ --- } \cdot \lambda, \; 2 \cdot 3 \cdot 4 \cdot \text{ --- } \cdot (\lambda+1), \; 3 \cdot 4 \cdot 5 \cdot \text{ --- } \cdot (\lambda+2),$$

$$\text{ --- } \cdot n(n+1)(n+2) \text{ --- } \cdot (n+\lambda-1) \text{ is the given sequence}$$

Theorem 1.1.

$$[\textstyle\coprod]_n \atop [\lambda] = {[\textstyle\coprod]_n^{[1]} \atop [\lambda]} \lambda \Big[a_1 + a_2(n-1) + a_3(n-2) + \text{ --- } + a_\lambda(n-(\lambda-1))$$

$$+ a_{\lambda+1}\Big[\frac{(n-\lambda+1)(n-\lambda)}{2} \Big] \Big]$$

where

$$a_1 = (\lambda-1)\,!, \quad a_2 = \lambda\,!, \quad a_3 = \frac{\lambda\,!\,(\lambda-1)}{2}, a_4 = \frac{\lambda\,!\,(\lambda-2)}{2}, \quad a_5 = \frac{\lambda\,!\,(\lambda-3)}{2}$$

$$\text{- - - - - - - - -}$$

$$a_\lambda = \frac{\lambda\,!\,(\lambda-(\lambda-2))}{2} = \lambda\,!, \quad a_{\lambda+1} = (\lambda-1)\,!. \tag{1.1}$$

$\Big\{$ *In* $[\coprod]_{n}^{[1]} \;=\; [\coprod]_{n}^{[1]},$ *"λ" stands for the Sequence-Level Of the "Nand Se-* $_{[\lambda]}$ $_{[\lambda]}$

quence", and the superscript (though not essential here) indicates the Super-Sum Sequence-Level of the Sequence. [**2**]

Proof. The Proof follows by the Traditional Method of Mathematical Induction on "n" for each λ-Level and on "λ" itself for the General Result.
For $\lambda = 2$ we have

$$[\coprod]_{n}^{[2]} = a_1 + a_2(n-1) + a_3\left[\frac{(n-1)(n-2)}{2}\right] = [\coprod]_{n}^{[1]}$$

$$a_1 = 1! = 1, \; a_2 = 2! = 2, \; a_3 = 2! \cdot \frac{1}{2} = 1$$

$$[\coprod]_{n}^{[2]} = 1\cdot 2, \; 2\cdot 3, \; 3\cdot 4, \; \cdots, \; n(n+1) = [\coprod]_{n}^{[1]}$$

$$[\coprod]_{n}^{[2]} = 2\left[1! + 2!(n-1) + \frac{2!\cdot 1!}{2}\left[\frac{(n-1)(n-2)}{2}\right]\right]$$

$$= 2\left[1 + 2(n-1) + \left[\frac{(n-1)(n-2)}{2}\right]\right] = [\coprod]_{n}^{[1]}$$

For $\lambda = 3$ we have

$$[\coprod]_{n}^{[3]} = a_1 + a_2(n-1) + a_3(n-2) + a_4\left[\frac{(n-2)(n-3)}{2}\right] = [\coprod]_{n}^{[1]}$$

$$a_1 = 2! = 2, \; a_2 = 3! = 6, \; a_3 = 3! \cdot \frac{2!}{2} = 3! = 6, \; a_4 = 2!$$

$$[\coprod]_{n}^{[3]} = 1\cdot 2\cdot 3, \; 2\cdot 3\cdot 4, \; 3\cdot 4\cdot 5, \; \cdots, \; n(n+1)(n+2)$$

$$= [\coprod]_{n}^{[1]}$$

$$[\amalg]_{\substack{n \\ [3]}} = 3\left[2! + 3!(n-1) + 3!(n-2) + 2!\left[\frac{(n-2)(n-3)}{2}\right]\right]$$

$$= 3\left[2 + 6(n-1) + 6(n-2) + 2\left[\frac{(n-2)(n-3)}{2}\right]\right]$$

$$= [\amalg]_{\substack{n \\ [3]}}^{[1]}$$

For $\lambda = 4$ we have

$$[\amalg]_{\substack{n \\ [4]}} = a_1 + a_2(n-1) + a_3(n-2) + a_4(n-3)$$

$$+ a_5\left[\frac{(n-3)(n-4)}{2}\right] = [\amalg]_{\substack{n \\ [4]}}^{[1]}$$

$$a_1 = 3! = 6, \quad a_2 = 4! = 24, \quad a_3 = 4! \cdot \frac{3!}{2} = 72,$$

$$a_4 = 4! \cdot \frac{2!}{2} = 72, \quad a_5 = 3! = 6$$

$$[\amalg]_{\substack{n \\ [4]}} = 1 \cdot 2 \cdot 3 \cdot 4, \quad 2 \cdot 3 \cdot 4 \cdot 5, \quad 3 \cdot 4 \cdot 5 \cdot 6, \quad \text{---},$$

$$n(n+1)(n+2)(n+3) = [\amalg]_{\substack{n \\ [4]}}^{[1]}$$

$$[\amalg]_{\substack{n \\ [4]}} = 4\left[3! + 4!(n-1) + 4! \cdot \frac{3!}{2}(n-2) + 4! \cdot \frac{2!}{2}(n-3)\right.$$

$$\left. + 3!\left[\frac{(n-3)(n-4)}{2}\right]\right]$$

$$= 4\left[6 + 24(n-1) + 72(n-2) + 72(n-3)\right.$$

$$\left. + 6\left[\frac{(n-3)(n-4)}{2}\right]\right] = [\amalg]_{\substack{n \\ [4]}}^{[1]}$$

For $\lambda = 5$ we have

$$[\amalg]_{n \atop [5]} = a_1 + a_2(n-1) + a_3(n-2) + a_4(n-3)$$

$$+ a_5(n-4) + a_6\left[\frac{(n-4)(n-5)}{2}\right] = [\amalg]_{n \atop [5]}^{[1]}$$

$$a_1 = 4! = 24, \quad a_2 = 5! = 120, \quad a_3 = 5! \cdot \frac{4!}{2} = 1440,$$

$$a_4 = 5! \cdot \frac{3!}{2} = 360, a_5 = 5! \cdot \frac{2!}{2} = 120, \quad a_6 = 4! = 24$$

$$[\amalg]_{n \atop [5]} = 1 \cdot 2 \cdot 3 \cdot 4 \cdot 5, \; 2 \cdot 3 \cdot 4 \cdot 5 \cdot 6, \; 3 \cdot 4 \cdot 5 \cdot 6 \cdot 7, \; ---,$$

$$n(n+1)(n+2)(n+3)(n+4) = [\amalg]_{n \atop [5]}^{[1]}$$

$$[\amalg]_{n \atop [5]} = 5\left[4! + 5!(n-1) + 5! \cdot \frac{4!}{2}(n-2) + 5! \cdot \frac{3!}{2}(n-3)\right.$$

$$\left. + 5! \cdot \frac{2!}{2}(n-4) + 4!\left[\frac{(n-3)(n-4)}{2}\right]\right]$$

$$= 4\left[24 + 120(n-1) + 1440(n-2) + 360(n-3)\right.$$

$$\left. + 120(n-4) + 24\left[\frac{(n-3)(n-4)}{2}\right]\right] = [\amalg]_{n \atop [5]}^{[1]}$$

$$>>>>>>>>>>>>>>>>>>>>>>>>>>>>$$

$$[\amalg]_{n \atop [\lambda]} = 1 \cdot 2 \cdot 3 \cdot \; --- \; \cdot \lambda, \; 2 \cdot 3 \cdot 4 \cdot \; --- \; \cdot (\lambda+1), \; 3 \cdot 4 \cdot 5 \; \; --- \; \cdot (\lambda+2),$$

$$--- \; ---, \; n(n+1)(n+2) \; --- \; \cdot (n+\lambda-1) \cdot [\amalg]_{n \atop [\lambda]}^{[1]}$$

$$[\amalg]_{n \atop [\lambda]} = \lambda\left[a_1 + a_2(n-1) + a_3(n-2) + \; --- \; + a_\lambda(n-(\lambda-1))\right.$$

$$\left. + a_{\lambda+1}\left[\frac{(n-\lambda+1)(n-\lambda)}{2}\right]\right] = [\amalg]_{n \atop [\lambda]}^{[1]}$$

$$\lambda\left[(\lambda-1)! + \lambda!(n-1) + \lambda!\frac{(\lambda-1)}{2}(n-2) + \; ---\right.$$

$$\left. + \lambda!\frac{(\lambda-(\lambda-2))}{2}(n-(\lambda-1)) + \right.$$

$$(\lambda - 1)! \left[\frac{(n - \lambda + 1)(n - \lambda)}{2} \right] \right] = \begin{matrix} [1] \\ [\text{Ш}]_n \\ [\lambda] \end{matrix}$$

$$\begin{matrix} [\text{Ш}]_n \\ [\lambda] \end{matrix} = n(n+1)(n+2) \cdots \cdot (n + \lambda - 1)$$

$$= \lambda \left[(\lambda - 1)! + \lambda!(n - 1) + \lambda! \frac{(\lambda - 1)}{2} (n - 2) + \cdots \right.$$

$$+ \lambda! \frac{(\lambda - (\lambda - 2))}{2} (n - (\lambda - 1)) +$$

$$(\lambda - 1)! \left[\frac{(n - \lambda + 1)(n - \lambda)}{2} \right] \right] = \begin{matrix} [1] \\ [\text{Ш}]_n \\ [\lambda] \end{matrix} \qquad \square$$

2. Formula for the Sum up to n Terms of the Nand Sequence

Let S_n be the Sum upto n terms of the Sequence

$$S_n \begin{bmatrix} [1] \\ [\text{Ш}] \\ [2] \end{bmatrix} = 1 \cdot 2 + 2 \cdot 3 + 3 \cdot 4 + \cdots + n(n+1)$$

$$\sum_{k=1}^{n} k(k+1) = \frac{n(n+1)(n+2)}{3} = 2! \left[\frac{n(n+1)(n+2)}{3!} \right] = \begin{matrix} [2] \\ [\text{Ш}]_n \\ [2] \end{matrix}$$

This is well known identity. [1]
The Following Identities can be Inductively Derived !

$$S_n \begin{bmatrix} [1] \\ [\text{Ш}] \\ [3] \end{bmatrix} = 1 \cdot 2 \cdot 3 + 2 \cdot 3 \cdot 4 + 3 \cdot 4 \cdot 5 + 6 \cdots$$

$$\sum_{k=1}^{n} \frac{k(k+1)(k+2)}{3} = \frac{n(n+1)(n+2)(n+3)}{4}$$

$$= 3! \left[\frac{n(n+1)(n+2)(n+3)}{4!} \right] = \begin{matrix} [2] \\ [\text{Ш}]_n \\ [3] \end{matrix}$$

$$S_n \begin{bmatrix} [1] \\ [\amalg] \\ [4] \end{bmatrix} = 1 \cdot 2 \cdot 3 \cdot 4 + 2 \cdot 3 \cdot 4 \cdot 5 + 3 \cdot 4 \cdot 5 \cdot 6 + \text{ --- } + n(n+1)(n+2)(n+3)$$

$$\sum_{k=1}^{n} \frac{k(k+1)(k+2)(k+3)}{4} = \frac{n(n+1)(n+2)(n+3)(n+4)}{5}$$

$$= 4! \left[\frac{n(n+1)(n+2)(n+3)(n+4)}{5!} \right]$$

$$= \begin{matrix} [2] \\ [\amalg]_n \\ [4] \end{matrix}$$

$$S_n \begin{bmatrix} [1] \\ [\amalg] \\ [5] \end{bmatrix} = 1 \cdot 2 \cdot 3 \cdot 4 \cdot 5 + 2 \cdot 3 \cdot 4 \cdot 5 \cdot 6 + 3 \cdot 4 \cdot 5 \cdot 6 \cdot 7$$
$$+ \text{ --- } + n(n+1)(n+2)(n+3)(n+4) =$$

$$\sum_{k=1}^{n} \frac{k(k+1)(k+2)(k+3)(k+4)}{5} =$$

$$\frac{n(n+1)(n+2)(n+3)(n+4)(n+5)}{6} =$$

$$5! \left[\frac{n(n+1)(n+2)(n+3)(n+4)(n+5)}{6!} \right] = \begin{matrix} [2] \\ [\amalg]_n \\ [5] \end{matrix}$$

$$- - - - - - - - - - - - - - - - - -$$

$$S_n \begin{bmatrix} [1] \\ [\amalg] \\ [\lambda] \end{bmatrix} = 1 \cdot 2 \cdot 3 \cdot \text{ --- } \cdot \lambda + 2 \cdot 3 \cdot 4 \cdot \text{ --- } \cdot (\lambda+1)$$
$$+ 3 \cdot 4 \cdot 5 \cdot \text{ --- } \cdot (\lambda+2) + \text{ --- --- }$$
$$+ n(n+1)(n+2) \text{ --- } (n+\lambda-1) =$$

$$\sum_{k=1}^{n} \frac{k(k+1)(k+2) \text{ --- } (k+\lambda-1)}{\lambda} =$$

$$\frac{n(n+1)(n+2) \text{ --- } (n+\lambda)}{(\lambda+1)} =$$

$$\lambda! \left[\frac{n(n+1)(n+2) \text{ --- } (n+\lambda)}{(\lambda+1)!} \right] = \begin{matrix} [2] \\ [\amalg]_n \\ [\lambda] \end{matrix} \qquad (2.1)$$

$$>>>>>>>>>>>>>>>>>>>>>>>>>>>>>$$

We can of course determine a Formula for the Super-Sum-α-Level by proceeding on inducting on α. Here $\alpha = 1$ and we have determined the Second-Level of the Given Sequence. [2, 3]

3. Another Equality

From (1.1) we have

$$
[\amalg]_n^{[1]} = [\amalg]_n^{} = \lambda\Big[a_1 + a_2(n-1) + a_3(n-2) + \, \cdots \\
[\lambda] \qquad [\lambda] \qquad\qquad + a_\lambda(n - (\lambda-1)) + a_{\lambda+1}\Big[\frac{(n-\lambda+1)(n-\lambda)}{2}\Big]\Big]
$$

$$
a_1 = (\lambda-1)!, \quad a_2 = \lambda!, \quad a_3 = \lambda!\frac{(\lambda-1)}{2}, \quad a_4 = \lambda!\frac{(\lambda-2)}{2},
$$

$$
a_5 = \lambda!\frac{(\lambda-3)}{2}
$$

$$
\cdots\cdots\cdots\cdots\cdots\cdots\cdots
$$

$$
a_\lambda = \lambda!\frac{(\lambda-(\lambda-2))}{2} = \lambda!, \quad a_{\lambda+1} = (\lambda-1)!
$$

Now we can easily prove by Induction that the sum of this progression up to n terms is

$$
S_n = [\amalg]^{[1]} = \lambda\Big[a_1 n + a_2\big[\frac{n(n-1)}{2}\big] + a_3\big[\frac{(n-1)(n-2)}{2}\big] + \, \cdots \\
[\lambda]
$$

$$
+ a_\lambda\big[\frac{(n-\lambda+1)(n-\lambda)}{2}\big] + a_{\lambda+1}\sum_1^{n-\lambda} i\frac{(n-i)(n-(i+1))}{2}\Big]
$$

$$
= \lambda\Big[a_1 n + a_2\big[\frac{n(n-1)}{2!}\big] + a_3\big[\frac{(n-1)(n-2)}{2!}\big] + \, \cdots \\
+ a_\lambda\big[\frac{(n-\lambda+1)(n-\lambda)}{2!}\big] + a_{\lambda+1}\big[\frac{(n-\lambda+2)(n-\lambda+1)(n-\lambda)}{3!}\big]\Big]
$$

$$
= \lambda\Big[(\lambda-1)![n] + \lambda!\big[\frac{n(n-1)}{2!}\big] + \lambda!\frac{(\lambda-1)}{2}\big[\frac{(n-1)(n-2)}{2!}\big] \\
+ \, \cdots + \lambda!\big[\frac{(n-\lambda+1)(n-\lambda)}{2!}\big] \\
+ (\lambda-1)!\big[\frac{(n-\lambda+2)(n-\lambda+1)(n-\lambda)}{3!}\big]\Big]
$$

$$= \lambda! \left[\frac{n(n+1)(n+2) \cdots (n+\lambda)}{(\lambda+1)!} \right] \quad [\text{ Substituting (2.1)}]$$

Dividing both sides by λ and simplifying we have

$$= \left[(\lambda-1)![n] + \lambda! [\frac{n(n-1)}{2!}] + \lambda! \frac{(\lambda-1)}{2} [\frac{(n-1)(n-2)}{2!}] \right.$$

$$+ \cdots + \lambda! [\frac{(n-\lambda+1)(n-\lambda)}{2!}]$$

$$\left. + (\lambda-1)! [\frac{(n-\lambda+2)(n-\lambda+1)(n-\lambda)}{3!}] \right]$$

$$= \left[\frac{n(n+1)(n+2) \cdots (n+\lambda)}{\lambda(\lambda+1)} \right] = \overset{[2]}{\underset{[\lambda]}{\coprod}}_n$$

We can of course determine a Formula for the Super-Sum-α-Level proceeding on inducting on α. Here $\alpha = 1$ and we have determined the Second-Level of the Given Sequence. [**2, 3**]

References

[1] David M. Burton: *ELEMENTARY NUMBER THEORY* UNIVERSAL BOOK STALL. NEW DELHI. Second Edition. Reprint 1998. (Ch.1)

[2] Narayanan Raghunathan: *Generalized Arithmetic Progressions – A Survey* [Submitted]

[3] Narayanan Raghunathan: *Functions and their Progressions – An Elementary Text*. [unpublished]

Rectangles, Golden Rectangle - A Brief Note

(Dedicated to Sri. CKUK - GPMHS - Palghat)

Abstract. In a Golden Rectangle (known since antiquity), if the largest square is cut away, then the figure remaining will also be a Golden Rectangle. Such Rectangles are characterized by a length-breadth ratio $\dfrac{1 + \sqrt{5}}{2}$, the Golden Ratio. This process clearly can be endlessly repeated. In this paper we note that for every Rectangle with a specific length-breadth ratio we can cut away a Rectangle with another characteristic length-breadth ratio such that the remaining Rectangle will have the same length-breadth ratio as the original ratio. This endlessly repeatable process clearly endows every Rectangle with a unique Extractable Sub-Rectangle (with a characteristic length-breadth ratio) associated with it. The formula to determine the length-breadth ratio of the Extractable Sub-rectangle from the length-breadth ratio of the original Rectangle and vice-versa are derived here. The Sum of the Areas of the Infinite Sequence of Rectangles of the same length-breadth ratio as the original Rectangle is also derived. The Golden Rectangle clearly is a special case when the extracted Sub Rectangle is a square!

Key Words:- Rectangles, Golden Rectangles, Generalization.
AMS Subject Classification No:- 11A99.

1. Infinite rectangles and their Raghunathan ratios [□]

We are clearly in the Euclidean-Terrain of Geometry. The ancient *terra-firma*! Let *ABCD* be any Rectangle. EVERY RECTANGLE is clearly characterized by the Fundamental Property that identifies it. ie. Its Length-Breadth Ratio [Basic Rectangular-Ratio].

$$\square = \frac{L}{B} = \frac{AB}{BC} > 1. \tag{1.1}$$

We are going to prove that we can extract INFINITE Rectangles of the same Length - Breadth Ratio from any arbitrary Rectangle.

We want $\dfrac{AB}{BC} = \dfrac{BC}{CE}$. Let $CE = k_1$. ie. $\dfrac{L}{B} = \dfrac{B}{k_1}$

So we have $k_1 = \dfrac{B^2}{L}$.

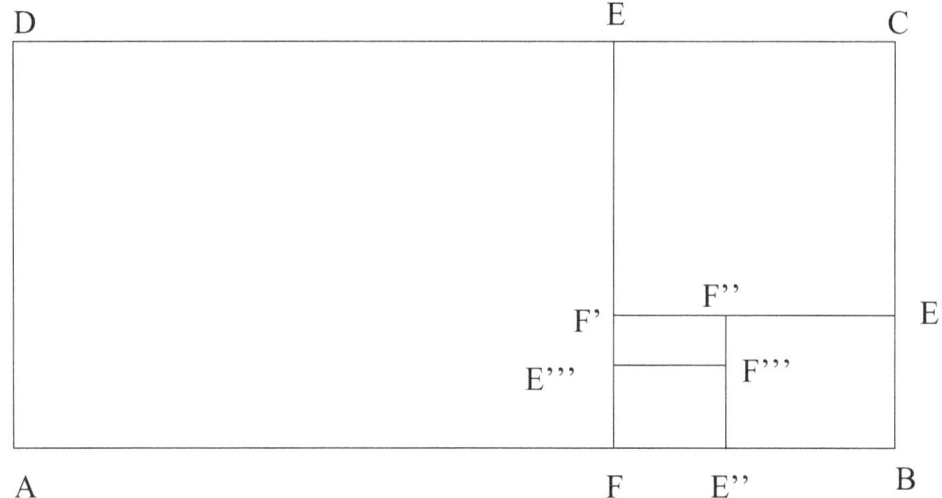

Proceeding on inductively we can prove similarly that

$$k_1 = \frac{B^2}{L}, k_2 = \frac{B^3}{L^2}, k_3 = \frac{B^4}{L^3}, \cdots, k_\tau = \frac{B^{(\tau+1)}}{L^\tau}$$

$$k_1 = CE, k_2 = BE', k_3 = FE'' \cdots \infty$$

(1.2)

The Length - Breadth Ratio of the Rectangle to be extracted from the original Rectangle is

$$\frac{AF}{AD} = \frac{(L - k_1)}{B} = \frac{(L - \frac{B^2}{L})}{B} = \frac{(L^2 - B^2)}{LB}$$

Dividing both the numerator and the denominator by B^2 and substituting

$$\square = \frac{L}{B} \text{ we have } \frac{(L^2 - B^2)}{LB} = \frac{(\square^2 - 1)}{\square}$$

For EVERY RECTANGLE we can clearly SEE that the AREA of the REC-TANGLE to be EXTRACTED from it to Retain Another RECTANGLE of the same RECTANGULAR RATIO [$\square$] has the RECTANGULAR RATIO given by the FORMULA

$$\overline{\square} = \frac{(L^2 - B^2)}{LB} = \frac{(L + B)(L - B)}{LB} = \frac{(\square^2 - 1)}{\square}$$

$$= \frac{(\square + 1)(\square - 1)}{\square}.$$

(1.3)

We will call $\overline{\square}$ THE RAGHUNATHAN RATIO of the RECTANGLE to HON-OUR my FATHER.
When THE RAGHUNATHAN RATIO is 1 ie.

$$\overline{\square} = \frac{(L^2 - B^2)}{LB} = 1 \text{ we have the Special Case of the so called Golden}$$

Rectangle when the RECTANGLE EXTRACTED is a SQUARE !

2. Formula to extract the basic rectangular ratio from the Raghunathan ratio

Let

$$\overline{\square} = \frac{(L^2 - B^2)}{LB}$$

Let $B = 1$ we have,

$$\frac{(L^2 - 1^2)}{L} = \overline{\square}$$
$$L^2 - 1 = \overline{\square}L \tag{2.1}$$
$$L^2 - \overline{\square}L - 1 = 0.$$

Solving the quadratic equation and taking the positive value of the solution we have

$$L = \frac{\overline{\square} + \sqrt{\overline{\square}^2 + 4}}{2}$$
$$\square = \frac{L}{B} = \frac{\overline{\square} + \sqrt{\overline{\square}^2 + 4}}{2}. \tag{2.2}$$

We will call this formula to determine the BASIC RECTANGULAR RATIO from the RAGHUNATHAN RATIO of THE RECTANGLE, THE RAJALAKSHMI FORMULA to HONOUR my MOTHER. When we substitutes $\overline{\square} = 1$ we have the Golden Rectangular Ratio

$$\square = \frac{1 + \sqrt{5}}{2}. \text{ (Ref: 1]-front coverpage. - 2]-p.136)}$$

$$1 < \square < \frac{1 + \sqrt{5}}{2} \iff 0 < \overline{\square} < 1 \text{ and } \square > \frac{1 + \sqrt{5}}{2} \iff \overline{\square} > 1$$

3. Formula for the sum of the areas of the infinite sequence of rectangles generated with the same rectangular ratio

Here we derive a formula for the SUM OF THE AREAS OF INFINITE SEQUENCE OF RECTANGLES GENERATED WITH THE SAME RECTANGULAR RATIO.

Let $\{ \overset{\square}{A_i} \}$ be the AREAS of the Rectangles

$$\overset{\square}{A_1}[ABCD] = LB, \ \overset{\square}{A_2}[FBCE] = \frac{B^3}{L}, \ \overset{\square}{A_3}[EBE'F'] = \frac{B^5}{L^3} \text{------}\infty$$

and let $[\overset{\square}{A}]$ denote the Sum of the Infinite Sequence of Areas.

$$[\overset{\square}{A}] = \sum_{i=1}^{\infty} \overset{\square}{A_i} = LB + \frac{B^3}{L} + \frac{B^5}{L^3} + \frac{B^7}{L^5} + \cdots$$

$$[\overset{\square}{A}] = LB + \frac{B^3}{L}\left[1 + \frac{B^2}{L^2} + \frac{B^4}{L^4} + \cdots\right].$$

(3.1)

Simplifying the Infinite Geometric Progression with the Common Ratio

$\dfrac{B^2}{L^2} < 1$, we have

$$[\overset{\square}{A}] = LB + \frac{B^3}{L}\left[\frac{1}{1 - \dfrac{B^2}{L^2}}\right]$$

$$[\overset{\square}{A}] = LB + \frac{B^3}{L}\left[\frac{L^2}{(L^2 - B^2)}\right] = LB + \left[\frac{LB^3}{(L^2 - B^2)}\right]$$

$$[\overset{\square}{A}] = LB\left[1 + \frac{B^2}{(L^2 - B^2)}\right] = \frac{LB(L^2)}{(L^2 - B^2)} = \frac{(L^3 B)}{(L^2 - B^2)}$$

$$= \frac{L^3 B}{(L + B)(L - B)} \tag{3.2}$$

$$[\overset{\square}{A}] = \frac{LB(L^2)}{(L^2 - B^2)} = \frac{L^2}{\square}\left[\frac{\square^2}{(\square^2 - 1)}\right] = \frac{L^2}{\square}\left[\frac{\square^2}{(\square + 1)(\square - 1)}\right] \tag{3.3}$$

$$[\overset{\square}{A}] = \frac{L^2}{\square}\left[\frac{\square^2}{(\square + 1)(\square - 1)}\right] = L^2\left[\frac{\square}{(\square + 1)(\square - 1)}\right] \tag{3.4}$$

$$[\overset{\square}{A}] = \frac{LB(L^2)}{(L^2 - B^2)} = B^2\square\left[\frac{\square^2}{(\square^2 - 1)}\right]$$

$$= B^2\square\left[\frac{\square^2}{(\square + 1)(\square - 1)}\right]. \tag{3.5}$$

We may also express $[\overset{\square}{A}]$ in terms of the Raghunathan Ratios. Substituting

$$\square = \frac{\overline{\square} + \sqrt{\overline{\square}^2 + 4}}{2} \; (2.2) \; \text{in} \; [\overset{\square}{A}] = \frac{L^2}{\square}\left[\frac{\square^2}{(\square^2 - 1)}\right] \; (3.3)$$

we have

$$[\overset{\square}{A}] = \left[\frac{L^2}{\left[\dfrac{\overline{\square} + \sqrt{\overline{\square}^2 + 4}}{2}\right]}\right]\left[\frac{\left[\dfrac{\overline{\square} + \sqrt{\overline{\square}^2 + 4}}{2}\right]^2}{\left[\dfrac{\overline{\square} + \sqrt{\overline{\square}^2 + 4}}{2}\right]^2 - 1}\right]$$

$$[\overset{\square}{A}] = \left[\cfrac{L^2}{\left[\cfrac{\overline{\square}+\sqrt{\overline{\square}^2+4}}{2}\right]}\right]\left[\cfrac{\cfrac{2\overline{\square}^2+2\overline{\square}\sqrt{\overline{\square}^2+4}+4}{4}}{\cfrac{2\overline{\square}^2+2\overline{\square}\sqrt{\overline{\square}^2+4}+4}{4}-1}\right]$$

$$[\overset{\square}{A}] = \left[\cfrac{L^2}{\left[\cfrac{\overline{\square}+\sqrt{\overline{\square}^2+4}}{2}\right]}\right]\left[\cfrac{2\overline{\square}^2+2\overline{\square}\sqrt{\overline{\square}^2+4}+4}{2\overline{\square}^2+2\overline{\square}\sqrt{\overline{\square}^2+4}}\right]$$

$$[\overset{\square}{A}] = \left[\cfrac{L^2}{\left[\cfrac{\overline{\square}+\sqrt{\overline{\square}^2+4}}{2}\right]}\right]\left[\cfrac{\overline{\square}^2+\overline{\square}\sqrt{\overline{\square}^2+4}+2}{\overline{\square}^2+\overline{\square}\sqrt{\overline{\square}^2+4}}\right]. \qquad (3.6)$$

Substituting

$$\square = \frac{\overline{\square}+\sqrt{\overline{\square}^2+4}}{2} \ (2.2) \text{ in } [\overset{\square}{A}] = L^2\left[\frac{\square}{(\square+1)(\square-1)}\right] \ (3.4)$$

we have

$$[\overset{\square}{A}] = L^2\left[\cfrac{\left[\cfrac{\overline{\square}+\sqrt{\overline{\square}^2+4}}{2}\right]}{\left[\cfrac{\overline{\square}+\sqrt{\overline{\square}^2+4}}{2}\right]^2-1}\right] = L^2\left[\cfrac{\overline{\square}+\sqrt{\overline{\square}^2+4}}{\overline{\square}^2+\overline{\square}\sqrt{\overline{\square}^2+4}}\right]. \qquad (3.7)$$

Substituting

$$\square = \frac{\overline{\square}+\sqrt{\overline{\square}^2+4}}{2} \ (2.2) \text{ in } [\overset{\square}{A}] = B^2\square\left[\frac{\square^2}{(\square+1)(\square-1)}\right] \ (3.5)$$

we have

$$[\overset{\square}{A}] = B^2\left[\frac{\overline{\square}+\sqrt{\overline{\square}^2+4}}{2}\right]\left[\frac{\overline{\square}^2+\overline{\square}\sqrt{\overline{\square}^2+4}+2}{\overline{\square}^2+\overline{\square}\sqrt{\overline{\square}^2+4}}\right]. \qquad (3.8)$$

The Formulae (3.2) - - - (3.8) may be called the Rajalakshmi-Raghunathan Formulae.

Sum of the Areas of the Rectangles Extracted

We may now consider the Area of the Sum of the Rectangles Extracted. We can intuitively see that it must be the Area of the Rectangle *ABCD* itself!

Let $\{ \bar{\Box}_{A_i} \}$ be the AREAS of the Rectangles Extracted. ie.

$$\bar{\Box}_{A_1}[AFED] = (LB - \frac{B^3}{L}),\ \bar{\Box}_{A_2}[F'E'CE] = (\frac{B^3}{L} - \frac{B^5}{L^3}),$$

$$\bar{\Box}_{A_3}[E''BE'F''] = (\frac{B^5}{L^3} - \frac{B^7}{L^5}) \text{------}\infty$$

and let $[\bar{\Box}_A]$ denote the Sum of the Infinite Sequence of Extracted Areas.

$$[\bar{\Box}_A] = \sum_{i=1}^{\infty} \bar{\Box}_{A_i} = (LB - \frac{B^3}{L}) + (\frac{B^3}{L} - \frac{B^5}{L^3}) + (\frac{B^5}{L^3} - \frac{B^7}{L^5}) + \cdots$$

$$[\bar{\Box}_A] = LB + (\frac{B^3}{L} + \frac{B^5}{L^3} + \frac{B^7}{L^5} + \cdots) - (\frac{B^3}{L} + \frac{B^5}{L^3} + \frac{B^7}{L^5} + \cdots) \qquad (3.9)$$

$$[\bar{\Box}_A] = LB.$$

Thus we may state an elegant (albeit trivial) Theorem.

Every Rectangle with the Rectangular Ratio $\Box = \dfrac{L}{B}$ may be Decomposed into an Infinite Family of Rectangles [Raghunathan-Rectangles] with the Rectangular Ratio $\bar{\Box} = \dfrac{(L+B)(L-B)}{LB} = \dfrac{(\Box+1)(\Box-1)}{\Box}$

We will conclude this paper with the observation that for every Rectangle with a certain Basic Rectangular Ratio there exists a unique Logarithmic Spiral associated with it. The Infinite Family of Raghunathan Rectangles of every Rectangle Circumscribe a Unique Logarithmic Spiral. When the Circumscribing Rectangle Finally Becomes A Point, the Spiral Culminates into the Source! Naturally!

I thank my brother Anand and friend Adrian for sharing my joy. I hesitantly told them, "Perhaps Pythagores, Heraclitus (!) or Aryabhatta has already written my paper milleniums ago". I apologize for my audacity of rescribing them, if these elementary results are already well-known!

Bibliography

1] George E. Andrews: *NUMBER THEORY*; HINDUSTAN PUBLISHING CORPORATION (INDIA) DELHI 110007. (1992)

2] Eli Maor: *e* : *THE STORY OF A NUMBER*; UNIVERSITIES PRESS (INDIA) LIMITED HYDERABAD (1999)

3] Narayanan Raghunathan : *INFINITE RECTANGLES AND THEIR LOGARITHMIC SPIRALS. A VISUAL MATHEMATICAL TREATISE*; [BEING PREPARED]

Matrix Properties of the Most Generalized Fibonacci Sequences

(Dedicated to my esteemed teacher Shri. K. M. Nair, who taught me at SSKZM)

Abstract. By using matrix properties of the most generalized Fibonacci sequences $\{o_n\}$, some properties and quadratic relations of the sequences $\{o_n\}$ are obtained; the sequences are defined by the General Infinite Family of recurrence relations:

$$o_n = o_{n-1} + o_{n-2} + o_{n-3} + \ldots + o_{n-(\zeta-1)} + o_{n-\zeta}, n \geq \zeta$$

$$[\zeta = 2, 3, \ldots \alpha].$$

Key Words:- Fibonacci Sequences, Infinite Generalizations.
AMS Subject Classification No:- 11B39; 11A99.

1. Inroduction

The Fibonacci sequence $\{F_n\}$ is defined as [1] $F_n = F_{n-1} + F_{n-2}$ and $F_0 = 0$, $F_1 = 1$. This sequence has been generalized in different ways.

(a) Modifying the recurrence relations such that each term is the sum of three preceding terms [2] i.e.

$$P_n = P_{n-1} + P_{n-2} + P_{n-3}$$

where P_0, P_1, P_2 are arbitrary algebraic integers all of which are not zero.

(b) Other modifying recurrence relations such that each term is the sum of four preceding terms [3], i.e.

$$Q_n = Q_{n-1} + Q_{n-2} + Q_{n-3} + Q_{n-4}$$

where Q_0, Q_1, Q_2, Q_3 are arbitrary algebraic integers all of which are not zero.

(c) Other modifying recurrence relations such that each term is the sum of five preceding terms [4], i.e.

$$D_n = D_{n-1} + D_{n-2} + D_{n-3} + D_{n-4} + D_{n-5}, \quad n \geq 5$$

where D_0, D_1, D_2, D_3, D_4 are arbitrary algebraic integers all of which are not zero.

Now, we shall generalize with the infinite family of recurrence relations in which each term is the sum of $\zeta(xi)$ $[\zeta = 2, 3, \ldots \alpha]$ preceding terms.

(This paper was published in the Journal of the Indian Academy of Mathematics)

2. The Generalized Sequences

$\{o_n\}$: We consider sequences:

$$\left\{o_n\right\} \equiv o_0,\ o_1,\ o_2,\ldots,o_n\ldots$$

where $o_0,\ o_1,\ o_2,\ o_3,\ldots,o_{\zeta-1}$, are arbitrary algebraic integers all of which are not zero and

$$o_n = o_{n-1} + o_{n-2} + o_{n-3} + \ldots + o_{n-(\zeta-1)} + o_{n-\zeta}, n \geq \zeta$$
$$[\zeta = 2, 3, \ldots \alpha]. \tag{2.1}$$

For $\zeta = 2$, it gives the basic family of Fibonacci sequences and for $o_0 = 0$, $o_1 = 1$, it gives the Fundamental Fibonacci sequence and the arguments of this paper are superfluous. For $\zeta = 3, 4, \ldots \alpha$, it is valid. Whenever, inductive arguments are used for the proof, there are two steps involved. First, prove it for $\zeta = 3, 4$ and 5 in each case proving by induction ! [$\zeta = 3$ is enough]. We assume that the concerned result is true for $\zeta = 3, 4, 5, \ldots, k$ and close the doubly inductive argument by proving it for $\zeta = k + 1$. The cumbersome and self-evident steps are omitted. { For, $\zeta = 5$ we get the results of Sanjay K. Harne and C. L. Parihar [4]}.

We also consider the sequences

$$\left\{{}^1\Delta_n\right\} \equiv {}^1\Delta_0, {}^1\Delta_1, {}^1\Delta_2, \ldots {}^1\Delta_n \ldots$$

where ${}^1\Delta_0 = o_2 - o_1 - o_0,\ {}^1\Delta_1 = o_3 - o_2 - o_1,\ {}^1\Delta_2 = o_4 - o_3 - o_2,\ldots$

$$^1\Delta_{(\zeta-2)} = o_\zeta - o_{\zeta-1} - o_{\zeta-2} \tag{2.2}$$

with

$$^1\Delta_n = o_{n-1} + o_{n-2} + o_{\zeta-2} + \ldots$$
$$+ o_{n-(\zeta-2)} + o_{n-(\zeta-1)} \quad n \geq \zeta - 1 \quad (2.3)$$

and

$$\left\{{}^2\Delta_n\right\} \equiv {}^2\Delta_0, {}^2\Delta_1, {}^2\Delta_2, \ldots {}^2\Delta_n \ldots$$
$$^2\Delta_0 = n_3 - o_2 - o_1 - o_0,\quad {}^2\Delta_1 = n_4 - o_3 - o_2 - o_1,$$
$$^2\Delta_2 = n_5 - o_4 - o_3 - o_2,$$
$$^2\Delta_{(\zeta-3)} = n_\zeta - o_{\zeta-1} - o_{\zeta-2} - o_{\zeta-3} \tag{2.4}$$

with

$$^2\Delta_n = o_{n-1} + o_{n-2} + \ldots$$
$$+ o_{n-(\zeta-3)} + o_{n-(\zeta-2)} n \geq \zeta - 2 \ldots \quad (2.5)$$
$$\left\{{}^{\zeta-3}\Delta_n\right\} \equiv {}^{\zeta-3}\Delta_0, {}^{\zeta-3}\Delta_1, {}^{\zeta-3}\Delta_2, \ldots {}^{\zeta-3}\Delta_n \ldots$$

where

$$\zeta^{-3}\Delta_0 = o_{\zeta-2} - o_{\zeta-3} - \ldots - o_1 - o_0$$
$$\zeta^{-3}\Delta_1 = o_{\zeta-1} - o_{\zeta-2} - \ldots - o_2 - o_1 \qquad (2.6)$$
$$\zeta^{-3}\Delta_2 = o_\zeta - o_{\zeta-1} - \ldots - o_3 - o_2$$

with

$$\zeta^{-3}\Delta_n = o_{n-1} + o_{n-2} + o_{n-3} \quad n \geq 3$$

and

$$\left\{ {}^{\zeta-2}\Delta_n \right\} \equiv {}^{\zeta-2}\Delta_0, {}^{\zeta-2}\Delta_1, {}^{\zeta-2}\Delta_2, \ldots {}^{\zeta-2}\Delta_n \ldots$$
$$\zeta^{-2}\Delta_0 = o_{\zeta-1} - o_{\zeta-2} - \ldots - o_1 - o_0 \qquad (2.7)$$
$$\zeta^{-2}\Delta_1 = o_{\zeta-2} - o_{\zeta-2} - \ldots - o_2 - o_1$$

with

$$\zeta^{-2}\Delta_0 = o_{n-1} + o_{n-2} \geq 2 \qquad (2.8)$$

From (2.3) and (2.1), wee have for $n \geq \zeta + (\zeta - 1)$

$$\begin{aligned}
{}^1\Delta_n &= o_{n-(\zeta+1)} + o_{n-\zeta} + o_{n-(\zeta-1)} + \ldots + o_{n-3} + o_{n-2} \\
&= o_{n-(\zeta+2)} + o_{n-(\zeta+1)} + o_{n-\zeta} + \ldots + o_{n-4} + o_{n-3} \\
&= o_{n-(\zeta+3)} + o_{n-(\zeta+2)} + o_{n-(\zeta+1)} + \ldots + o_{n-5} + o_{n-4} \ldots \\
&= o_{n-[\zeta+(\zeta-1)]} + o_{n-[\zeta+(\zeta-2)]} + o_{n-[\zeta+(\zeta-3)]} + \ldots \\
&\quad + o_{n-(\zeta+1)} + o_{n-\zeta} \\
{}^1\Delta_n &= {}^1\Delta_{n-1} + {}^1\Delta_{n-2} + {}^1\Delta_{n-3} + \ldots + {}^1\Delta_{n-(\zeta-1)} + {}^1\Delta_{n-\zeta}.
\end{aligned}$$

Using (2.3) and (2.2), we obtain:

$$\begin{aligned}
{}^1\Delta_{\zeta+(\zeta-2)} &= \left(o_{\zeta+1} + o_\zeta + o_{\zeta-1} + o_{\zeta-2} \right) \\
&\quad + \left(o_\zeta + o_{\zeta-1}2 + o_{\zeta-2} + o_{\zeta-3} \right) \ldots + \left(o_4 + o_3 + o_2 + o_1 \right) \\
&\quad + \left(o_3 + o_2 + o_1 + o_0 \right) + \left(o_\zeta - o_{\zeta-1} - o_{\zeta-2} \right) \\
{}^1\Delta_{\zeta+(\zeta-2)} &= {}^1\Delta_{\zeta+(\zeta-3)} + {}^1\Delta_{\zeta+(\zeta-4)} + \ldots + {}^1\Delta_\zeta + {}^1\Delta_{\zeta-1} + {}^1\Delta_{\zeta-2}
\end{aligned}$$

and similarly,

$$\begin{aligned}
{}^1\Delta_{\zeta+(\zeta-3)} &= {}^1\Delta_{\zeta+(\zeta-4)} + {}^1\Delta_{\zeta+(\zeta-5)} + \ldots + {}^1\Delta_{\zeta-1} + {}^1\Delta_{\zeta-2} + {}^1\Delta_{\zeta-3} \\
{}^1\Delta_{\zeta+(\zeta-4)} &= {}^1\Delta_{\zeta+(\zeta-5)} + {}^1\Delta_{\zeta+(\zeta-6)} + \ldots + {}^1\Delta_{\zeta-2} + {}^1\Delta_{\zeta-3} + {}^1\Delta_{\zeta-4} \\
{}^1\Delta_{\zeta+1} &= {}^1\Delta_\zeta + {}^1\Delta_{\zeta-1} + \ldots + {}^1\Delta_3 + {}^1\Delta_2 + {}^1\Delta_1.
\end{aligned}$$

Hence, we have for $n \geq \zeta$

$$^1\Delta_n = {}^1\Delta_{n-1} + {}^1\Delta_{n-2} + {}^1\Delta_{n-3} + \ldots + {}^1\Delta_{n-(\zeta-1)} + {}^1\Delta_{n-\zeta}. \qquad (2.9)$$

Proceeding on similar lines, it can be easily shown that for $n \geq \zeta$.

$$^2\Delta_n = o_{n-2} + o_{n-3} + \ldots + o_{n-\zeta} + o_{n-(\zeta+1)}$$
$$+ o_{n-3} + o_{n-4} + \ldots + o_{n-(\zeta+1)} + o_{n-(\zeta+2)} \cdots$$
$$+ o_{n-(\zeta-1)} + o_{n-\zeta} + \ldots + o_{n-(\zeta+\zeta-3)} + o_{n-(\zeta+\zeta-2)}$$
$$^2\Delta_n = {}^2\Delta_{n-1} + {}^2\Delta_{n-2} + \ldots + {}^2\Delta_{n-(\zeta-1)} + {}^2\Delta_{n-\zeta} \text{ for } n \geq \zeta.$$

(2.10)

Proceeding on similar lines, it can be easily shown that for $n \geq \zeta$ [$\zeta = 3, 4, 5, \ldots \alpha$] and for $^3\Delta_n$, $^4\Delta_n, \ldots$ similar relations hold finally culminating in $^{\zeta-2}\Delta_n$ such that

$$^{\zeta-2}\Delta_n = o_{n-2} + o_{n-3} + \ldots + o_{n-\zeta} + o_{n-(\zeta+1)}$$
$$+ o_{n-3} + o_{n-4} + \ldots + o_{n-(\zeta+1)} + o_{n-(\zeta+2)}$$
$$^{\zeta-2}\Delta_n = {}^{\zeta-2}\Delta_{n-1} + {}^{\zeta-2}\Delta_{n-2} + \ldots$$
$$+ {}^{\zeta-2}\Delta_{n-(\zeta-1)} + {}^{\zeta-2}\Delta_{n-\zeta} \text{ for } n \geq \zeta.$$

(2.11)

Thus, the $(\zeta - 2)$ sequences $\{^3\Delta_n\}$, $\{^2\Delta_n\}, \ldots \{^{\zeta-3}\Delta_n\}$ and $\{^{\zeta-2}\Delta_n\}$ are special cases for sequences $\{o_n\}$ and are obtained by taking different initial values.

$$o_n = o_{n-1} + o_{n-2} + o_{n-3} + \ldots + o_{n-(\zeta-1)} + o_{n-\zeta}, n \geq \zeta$$
$$[\zeta = 2, 3, \ldots \alpha].$$

(2.1)

on taking

$$o_0 = o_1 = \ldots = o_{\zeta-4} = 0; \ o_{\zeta-3} = 1; \ o_{\zeta-2} = 1; \ o_{\zeta-1} = 2$$
$$o_0 = o_1 = \ldots = o_{\zeta-5} = 0; \ o_{\zeta-4} = 1; \ o_{\zeta-3} = 0; o_{\zeta-2} = 1; \ o_{\zeta-1} = 2$$
$$o_0 = o_1 = \ldots = o_{\zeta-6} = 0; \ o_{\zeta-5} = 1; \ o_{\zeta-4} = o_{\zeta-3} = 0; \ o_{\zeta-2} = 1; \ o_{\zeta-1} = 2$$
$$o_0 = o_1 = \ldots = o_{\zeta-7} = 0; \ o_{\zeta-6} = 1; \ o_{\zeta-5} = \ldots = o_{\zeta-3} = 0;$$
$$o_{\zeta-2} = 1; \ o_{\zeta-1} = 2 \ldots$$
$$o_0 = o_1 = 0; \ o_2 = 1; \ o_3 \ldots = o_{\zeta-3} = 0; \ o_{\zeta-2} = 1; \ o_{\zeta-1} = 2$$
$$o_0 = 0; \ o_1 = 1; \ o_2 = \ldots = o_{\zeta-3} = 0; \ o_{\zeta-2} = 1; \ o_{\zeta-1} = 2$$
$$o_0 = 1; \ o_1 = o_2 = \ldots = o_{\zeta-3} = 0; \ o_{\zeta-2} = 1; \ o_{\zeta-1} = 2$$
$$o_0 = o_1 = o_2 = o_3 \ldots o_{\zeta-3} = 0; \ o_{\zeta-2} = 1; \ o_{\zeta-1} = 2$$

$$0, 0, \ldots 0, 0, 1, 1, 2, \ldots, {}^1A_n, \ldots 0, 0, \ldots 0, 1, 0, 1, 2, \ldots, {}^2A_n, \ldots$$
$$0, 0, \ldots 1, 0, 0, 1, 2, \ldots, {}^3A_n, \ldots 0, 0, \ldots 1, 0, 0, 0, 1, 2, \ldots, {}^4A_n, \ldots$$
$$\cdots\cdots\cdots\cdots\cdots\cdots\cdots\cdots\cdots$$
$$0, 0, 1, \ldots 0, 1, 2, \ldots, {}^{\zeta-4}A_n, \ldots 0, 1, \ldots 0, 1, 2, \ldots, {}^{\zeta-3}A_n, \ldots$$
$$1, 0, \ldots 0, 1, 2, \ldots, {}^{\zeta-2}A_n, \ldots 0, 0, \ldots 0, 1, 2, \ldots, {}^{\zeta-1}A_n, \ldots$$

(2.12)

Here, we find that

$$
\begin{aligned}
{}^{2}A_n &= {}^{1}A_{n-1} +{}^{1} A_{n-2} +{}^{1} A_{n-3} + \ldots {}^{1}A_{n-(\zeta-1)} \\
{}^{3}A_n &= {}^{1}A_{n-1} +{}^{1} A_{n-2} +{}^{1} A_{n-3} + \ldots {}^{1}A_{n-(\zeta-2)} \\
&\cdots\cdots\cdots\cdots\cdots\cdots\cdots\cdots\cdots\cdots\cdots\cdots \\
{}^{\zeta-2}A_n &= {}^{1}A_{n-1} +{}^{1} A_{n-2} +{}^{1} A_{n-3} \\
{}^{\zeta-1}A_n &= {}^{1}A_{n-1} +{}^{1} A_{n-2}.
\end{aligned}
$$

Thus, we can say that $({}^{1}\Delta_n)$ is a o_n-type sequence while $({}^{2}\Delta_n)$ is a $({}^{1}\Delta_n)$-type sequence etc. . . and $({}^{\zeta-2}\Delta_n)$ is a ${}^{\zeta-3}\Delta_n$-type sequence and $({}^{\zeta-1}\Delta_n)$ is a ${}^{\zeta-2}\Delta_n$-type sequence.

3. Simple Properties

Let us consider the $[\zeta \times \zeta]$ matrices where $\zeta = 3, 4, 5, \ldots \alpha$

$$
X \equiv
\begin{bmatrix}
1 & 1 & \ldots & 1 & 1 & 1 \\
1 & 0 & \ldots & 0 & 0 & 0 \\
0 & 1 & \ldots & 0 & 0 & 0 \\
\multicolumn{6}{c}{\cdots\cdots\cdots\cdots\cdots\cdots} \\
0 & 0 & \ldots & 1 & 0 & 0 \\
0 & 0 & \ldots & 0 & 1 & 0
\end{bmatrix}.
\tag{3.1}
$$

By repeated mathematical induction it is easily proved that:

$$
[X]^n =
\begin{bmatrix}
{}^{1}A_{n+1} & {}^{2}A_{n+1} & \cdots & {}^{\zeta-2}A_{n+1} & {}^{\zeta-1}A_{n+1} & {}^{1}A_n \\
{}^{1}A_n & {}^{2}A_n & \cdots & {}^{\zeta-2}A_n & {}^{\zeta-1}A_n & {}^{1}A_{n-1} \\
{}^{1}A_{n-1} & {}^{2}A_{n-1} & \cdots & {}^{\zeta-2}A_{n-1} & {}^{\zeta-1}A_{n-1} & {}^{1}A_{n-2} \\
\multicolumn{6}{c}{\cdots\cdots\cdots\cdots\cdots\cdots\cdots\cdots\cdots\cdots\cdots\cdots} \\
{}^{1}A_{n-(\zeta-3)} & {}^{2}A_{n-(\zeta-3)} & \cdots & {}^{\zeta-2}A_{n-(\zeta-3)} & {}^{\zeta-1}A_{n-(\zeta-3)} & {}^{1}A_{n-(\zeta-2)} \\
{}^{1}A_{n-(\zeta-2)} & {}^{2}A_{n-(\zeta-2)} & \cdots & {}^{\zeta-2}A_{n-(\zeta-2)} & {}^{\zeta-1}A_{n-(\zeta-2)} & {}^{1}A_{n-(\zeta-1)}
\end{bmatrix}
$$
$$
n \geq \zeta - 1 \tag{3.2}
$$

and

$$
\left[o_n, o_{n-1}, o_{n-2}, \ldots, o_{n-(\zeta-2)}, o_{n-(\zeta-1)} \right]
$$
$$
= X^{n-(\zeta-1)} \left[o_{(\zeta-1)}, o_{(\zeta-2)}, \ldots, n_2, o_1, o_0 \right], \quad n \geq \zeta - 1. \tag{3.3}
$$

On using (3.2) and (3.3) we get:

$$
\begin{bmatrix} o_{n+p} \\ o_{n+p-1} \\ o_{n+p-2} \\ \cdots \\ o_{n+p-(\zeta-2)} \\ o_{n+p-(\zeta-1)} \end{bmatrix}
$$

$$
= \begin{bmatrix} {}^{1}A_{n+1} & {}^{2}A_{n+1} & \cdots & {}^{\zeta-2}A_{n+1} & {}^{\zeta-1}A_{n+1} & {}^{1}A_{n} \\ {}^{1}A_{n} & {}^{2}A_{n} & \cdots & {}^{\zeta-2}A_{n} & {}^{\zeta-1}A_{n} & {}^{1}A_{n-1} \\ {}^{1}A_{n-1} & {}^{2}A_{n-1} & \cdots & {}^{\zeta-2}A_{n-1} & {}^{\zeta-1}A_{n-1} & {}^{1}A_{n-2} \\ \cdots & \cdots & \cdots & \cdots & \cdots & \cdots \\ {}^{1}A_{n-(\zeta-3)} & {}^{2}A_{n-(\zeta-3)} & \cdots & {}^{\zeta-2}A_{n-(\zeta-3)} & {}^{\zeta-1}A_{n-(\zeta-3)} & {}^{1}A_{n-(\zeta-2)} \\ {}^{1}A_{n-(\zeta-2)} & {}^{2}A_{n-(\zeta-2)} & \cdots & {}^{\zeta-2}A_{n-(\zeta-2)} & {}^{\zeta-1}A_{n-(\zeta-2)} & {}^{1}A_{n-(\zeta-1)} \end{bmatrix} \begin{bmatrix} o_{n} \\ o_{n-1} \\ o_{n-2} \\ \cdots \\ o_{n-(\zeta-2)} \\ o_{n-(\zeta-1)} \end{bmatrix}.
$$

From this we obtain:

$$
o_{n+p} = {}^{1}A_{p+1}o_{n} + {}^{2}A_{p+1}o_{n-1} + \ldots
$$
$$
+ {}^{\zeta-2}A_{p+1}o_{n-(\zeta-3)} + {}^{\zeta-1}A_{p+1}o_{n-(\zeta-2)} + {}^{1}A_{n}o_{n-(\zeta-1)}. \quad (3.4)
$$

Let us consider the matrices $[Y]$ which are the transposes of the matrices $[X]$, i.e.,

$$
[Y] = [X] \begin{bmatrix} 1 & 1 & 0 & 0 & \ldots & 0 \\ 1 & 0 & 1 & 0 & \ldots & 0 \\ 1 & 0 & 0 & 1 & \ldots & 0 \\ \cdots & \cdots & \cdots & \cdots & \cdots & \cdots \\ 1 & 0 & 0 & 0 & \ldots & 1 \\ 1 & 0 & 0 & 0 & \ldots & 0 \end{bmatrix}.
$$

It can be shown easily that the sequences:

$$
o_{3}, {}^{1}\Delta_{4}, \ldots {}^{\zeta-3}\Delta_{4}, {}^{\zeta-2}\Delta_{4}, o_{4}, \ldots o_{n-1}, {}^{1}\Delta_{n}, \ldots {}^{\zeta-3}\Delta_{n}, {}^{\zeta-2}\Delta_{n}, o_{n} \quad (3.5)
$$

are generalized by the matrices $[Y]$, that is,

$$
\left[o_{n}, {}^{1}\Delta_{n}, \ldots {}^{\zeta-3}\Delta_{n}, {}^{\zeta-2}\Delta_{n}, o_{n-1} \right]
$$
$$
= X^{n-(\zeta-1)} \left[o_{\zeta-1}, {}^{1}\Delta_{\zeta-1}, \ldots {}^{\zeta-3}\Delta_{\zeta-1}, {}^{\zeta-2}\Delta_{\zeta-1}, o_{\zeta-2} \right] n \geq \zeta - 1. \quad (3.6)
$$

On using (3.6) and (3.5), we get:

$$\left[o_{n+p},\, {}^{1}\Delta_{n+p}, \ldots\, {}^{\zeta-3}\Delta_{n+p},\, {}^{\zeta-2}\Delta_{n+p},\, {}^{1}A_{n+p-1} \right]$$

$$= y^{n+p-(\eta-1)} \left[o_{\zeta-1},\, {}^{1}\Delta_{\zeta-1}, \ldots\, {}^{\zeta-3}\Delta_{\zeta-1},\, {}^{\zeta-2}\Delta_{\zeta-1},\, o_{\zeta-2} \right]$$

$$= [Y]^{p} \left[o_{n},\, {}^{1}\Delta_{n}, \ldots\, {}^{\zeta-3}\Delta_{n},\, {}^{\zeta-2}\Delta_{n},\, o_{\zeta-1} \right]$$

$$= \begin{bmatrix} {}^{1}A_{p+1} & {}^{1}A_{p} & {}^{1}A_{p-1} & \cdots & {}^{1}A_{p-(\zeta-3)} & {}^{1}A_{p-(\zeta-2)} \\ {}^{2}A_{p+1} & {}^{2}A_{p} & {}^{2}A_{p-1} & \cdots & {}^{2}A_{p-(\zeta-3)} & {}^{2}A_{p-(\zeta-2)} \\ \cdots\cdots\cdots\cdots\cdots\cdots\cdots\cdots\cdots\cdots\cdots\cdots\cdots\cdots\cdots \\ {}^{\zeta-2}A_{p+1} & {}^{\zeta-2}A_{p} & {}^{\zeta-2}A_{p-1} & \cdots & {}^{\zeta-2}A_{p-(\zeta-3)} & {}^{\zeta-2}A_{p-(\zeta-2)} \\ {}^{\zeta-1}A_{p+1} & {}^{\zeta-1}A_{p} & {}^{\zeta-1}A_{p-1} & \cdots & {}^{\zeta-1}A_{p-(\zeta-3)} & {}^{\zeta-1}A_{p-(\zeta-2)} \\ {}^{1}A_{p} & {}^{1}A_{p-1} & {}^{1}A_{p-2} & \cdots & {}^{1}A_{p-(\zeta-2)} & {}^{1}A_{p-(\zeta-1)} \end{bmatrix} \begin{bmatrix} o_{n} \\ {}^{1}\Delta_{n} \\ \cdots \\ {}^{\zeta-3}\Delta_{n} \\ {}^{\zeta-2}\Delta_{n} \\ o_{n-1} \end{bmatrix}$$

Thus

$$o_{n+p} = {}^{1}A_{p+1} + {}^{1}A_{p}\,{}^{1}\Delta_{n} + {}^{1}A_{p-1}\,{}^{2}\Delta_{n} \ldots$$

$$+ {}^{1}A_{p-(\zeta-3)}\,{}^{\zeta-2}\Delta_{n} + {}^{1}A_{p-(\zeta-2)}o_{n-1}$$

$${}^{1}\Delta_{n+p} = {}^{2}A_{p+1} + {}^{2}A_{p}\,{}^{1}\Delta_{n} + {}^{2}A_{p-1}\,{}^{2}\Delta_{n} \ldots$$

$$+ {}^{2}A_{p-(\zeta-3)}\,{}^{\zeta-2}\Delta_{n} + {}^{2}A_{p-(\zeta-2)}o_{n-1}$$

$$\cdots\cdots\cdots\cdots\cdots\cdots\cdots\cdots\cdots\cdots\cdots\cdots\cdots$$

$${}^{\zeta-3}\Delta_{n+p} = {}^{\zeta-2}A_{p+1}o_{n} + {}^{\zeta-2}A_{p}\,{}^{1}\Delta_{n} + {}^{\zeta-2}A_{p-1}\,{}^{2}\Delta_{n} \ldots$$

$$+ {}^{\zeta-2}A_{p-(\zeta-3)}^{\zeta-2}\Delta_{n} + {}^{\zeta-2}A_{p-(\zeta-2)}o_{n-1}$$

$${}^{\zeta-2}\Delta_{n+p} = {}^{\zeta-1}A_{p+1}o_{n} + {}^{\zeta-1}A_{p}\,{}^{1}\Delta_{n} + {}^{\zeta-1}A_{p-1}\,{}^{2}\Delta_{n} \ldots$$

$$+ {}^{\zeta-1}A_{p-(\zeta-3)}^{\zeta-2}\Delta_{n} + {}^{\zeta-1}A_{p-(\zeta-2)}o_{n-1}.$$

4. Quadratic Relations

It is easily by seen by Repeated Induction that for $n \geq (\zeta - 1)$

$$\left[\zeta = 3, 4, 5, \ldots \alpha \right] \left[o_{n}, o_{n-1},\, o_{n-2}, \ldots, o_{n-(\zeta-2)}, o_{n-(\zeta-1)} \right]$$

$$= \left[o_{(\zeta-1)}, o_{(\zeta-2)}, \ldots, o_{2}, o_{1}, o_{0} \right], \quad [Y]^{n-1} \tag{4.1}$$

and

$$\left[o_{n},\, {}^{1}\Delta_{n}, \ldots\, {}^{\zeta-3}\Delta_{n},\, {}^{\zeta-2}\Delta_{n}, o_{n-1} \right]$$

$$= \left[o_{\zeta-1},\, {}^{1}\Delta_{\zeta-1}, \ldots\, {}^{\zeta-3}\Delta_{\zeta-1},\, {}^{\zeta-2}\Delta_{\zeta-1}o_{\zeta-2} \right] [X]^{n-(\zeta-1)}. \tag{4.2}$$

We shall now use the vector matrix representations of $o_{r}s$ to prove the following family of relations:

$$o_n^2 + o_{n-1}^2 + o_{n-2}^2 + \ldots + o_{n-(\zeta-2)}^2$$

$$+ 2o_{n-1}\Big\{o_{n-2} + \ldots + o_{n-(\zeta-2)} + o_{n-(\zeta-1)}\Big\} \tag{4.3}$$

$$= o_{\zeta-1}o_{2n-(\zeta-1)} + o_{\zeta-2}{}^1\Delta_{2n-(\zeta-1)} \ldots$$

$$+ o_2{}^{\zeta-3}\Delta_{2n-(\zeta-1)} + o_1{}^{\zeta-2}\Delta_{2n-(\zeta-1)} + o_0 o_{2n-\zeta}$$

and also

$$= o_{\zeta-1}o_{2n-(\zeta-1)} + {}^1\Delta_{\zeta-1}o_{2n-\zeta} + \ldots + {}^{\zeta-3}\Delta_{\zeta-1}o_{2n-(2\zeta-4)}$$

$$+ {}^{\zeta-2}\Delta_{\zeta-1}o_{2n-(2\zeta-3)} + o_{\zeta-2}o_{2n-(2\zeta-2)}. \tag{4.4}$$

Proof. The left hand sides are the scalar products of the vectors.

$$\Big[o_n, o_{n-1}, o_{n-2}, \ldots, o_{n-(\zeta-2)}, o_{n-(\zeta-1)}\Big]$$

and

$$\Big[o_n, {}^1\Delta_n, \ldots, {}^{\zeta-3}\Delta_n, {}^{\zeta-2}\Delta_n, o_{n-1}\Big].$$

Putting the value of equation (4.1) in the left hand side of equation (4.3), we get:

$$\Big[o_{\zeta-1}, o_{n-2}, \ldots, o_2, o_1, o_0\Big]X^{n-4}$$

$$\Big[o_n, {}^1\Delta_n, \ldots, {}^{\zeta-3}\Delta_n, {}^{\zeta-2}\Delta_n, o_{n-1}\Big].$$

Now, using (3.6) in the above equations, we get:

$$\Big[o_{\zeta-1}, o_{n-2}, \ldots, o_2, o_1, o_0\Big]$$

$$\Big[o_{2n-(\zeta-1)}, {}^1\Delta_{2n-(\zeta-1)}, \ldots, {}^{\zeta-3}\Delta_{2n-(\zeta-1)}, {}^{\zeta-2}\Delta_{2n-(\zeta-1)}, o_{2n-1}\Big]$$

which yields the R. H. S. of equation (4.3). The second set of relations (4.4) can also be similarly proved because:

$$\Big[o_n, {}^1\Delta_n, \ldots, {}^{\zeta-3}\Delta_n, {}^{\zeta-2}\Delta_n, o_{n-1}\Big]$$

$$\Big[o_n, o_{n-1}, o_{n-2}, \ldots, o_{n-(\zeta-2)}, o_{n-(\zeta-1)}\Big]$$

$$= \Big[o_{\zeta-1}, {}^1\Delta_{\zeta-1}, \ldots, {}^{\zeta-3}\Delta_{\zeta-1}, {}^{\zeta-2}\Delta_{\zeta-1}, o_{\zeta-2}\Big]Y^{n-(\zeta-1)}$$

$$\Big[o_n, o_{n-1}, o_{n-2}, \ldots, o_{n-(\zeta-2)}, o_{n-(\zeta-1)}\Big]$$

$$= \Big[o_{\zeta-1}, {}^1\Delta_{\zeta-1}, \ldots, {}^{\zeta-3}\Delta_{\zeta-1}, {}^{\zeta-2}\Delta_{\zeta-1}, o_{\zeta-2}\Big]$$

$$\Big[o_{2n-(\zeta-1)}, o_{2n-\zeta}, \ldots, o_{2n-(2\zeta-4)}, o_{2n-(2\zeta-3)}, o_{2n-(\zeta-2)}\Big]$$

$$= o_{\zeta-1}o_{2n-(\zeta-1)} + {}^1\Delta_{\zeta-1}o_{2n-\zeta} + \ldots + {}^{\zeta-3}\Delta_{\zeta-1}o_{2n-(2\zeta-4)}$$

$$+ {}^{\zeta-2}\Delta_{\zeta-1}o_{2n-(2\zeta-3)} + o_{\zeta-2}o_{2n-(2\zeta-2)}$$

$$= R.H.S. \qquad\qquad \square$$

Before I conclude, I must confess that it is the sheer Symmetry that leads me to this generalized paper. If we substitute $\zeta = 5$, we get Sanjay K. Harne and C. L. Parihar's results. We can clearly recognize now, that for each value of $\zeta \geq 3$ $[\zeta = 3, 4, \ldots, 6, 7, \ldots, \alpha]$, we get exactly similar symmetric results. Surely, strange is this vast magic of Fibonacci symmetry.

References

[1] Brother, U. Alfred: *An introduction to Fibonacci discovery*, The Fibonacci Association, Calif., (1965).

[2] Waddill and Lovis Sacks: *Fibonacci Quarterly*, 5(3), (1967), 209–222.

[3] Jaiswal D. V.: *LABDEV Jour. of Sequences and Technology (India)*, 7(2), (1969), 67–71.

[4] Sanjay Harne and C. L. Parihar: Matrix properties of generalized Fibonacci sequence, *J. Indian Acad. Math.*. Vol.17, No.2, 1995, 171–177.

Infinite Family of Generalized Fibonacci Polynomials

(Dedicated to Mrs. Jessie John, Ex. Matron, SSKZM)

Abstract. In this brief note, Fibonacci Polynomials are generalized by:

$$H_0(x) = 0, \ H_1(x) = 1, \ H_2(x) = a_{\xi-2}x,$$
$$H_3(x) = a_{\xi-3}x, \ \ldots, \ H_{\xi-1}(x) = a_1 x$$

and for $n \geq 0$

$$H_{n+\xi}(x) = a_1 x H_{n+\xi-1}(x) + a_2 x H_{n+\xi-2}(x) + \ldots$$
$$+ a_{\xi-3} x H_{n+3}(x) + a_{\xi-2} x H_{n+2}(x)$$
$$+ a_{\xi-1} H_{n+1}(x) + a_\xi H_n(x).$$

where $a_1, a_2, a_3 \ldots a_{\xi-1}, a_\xi$ are arbitrary algebraic integers and $\xi = 2, 3, \ldots \infty$.

Note this notation $H_n^{[\xi]}(x)$ where ξ may be called the level of the Generalized Fibonacci Polynomials.

Key Words:- Fibonacci Polynomials, Infinite Generalizations.
AMS Subject Classification No:- 11B39; 11A99.

1. Introduction

The Fibonacci Polynomials $f_1(x)$ [1, 2] is defined by :

$$F_0(x) = 0 f_1(x) = 1 \text{ and } f_{n+2}(x) = x f_{n+1}(x) + f_n(x). \tag{1.1}$$

and Jaiswal Polynomials $u_n(x)$ [3] defined by :

$$u_0(x) = 0 u_1(x) = 1 u_2(x) = ax \text{ and for } n \geq 0$$
$$u_{n+3}(x) = ax u_{n+2}(x) + b u_{n+1}(x) + c u_n(x) \tag{1.2}$$

where a, b, c are arbitrary algebraic integers and Harne and Parihar Polynomials $h_n(x)$ [4] defined by

$$h_0(x) = 0 h_1(x) = 1 h_2(x) = bx h_3(x) = ax \text{ and for } n \geq 0$$
$$h_{n+4}(x) = ax h_{n+3}(x) + bx h_{n+2}(x) + c h_{n+1}(x) + d h_n(x) \tag{1.3}$$

where a, b, c and d are arbitrary algebraic integers.

(This paper was published in the Journal of the Indian Academy of Mathematics)

2. General Family of Infinite Fibonacci Polynomials $H_n^{[\xi]}(x)$

$$H_0(x) = 0, \ H_1(x) = 1, \ H_2(x) = a_{\xi-2}x,$$
$$H_3(x) = a_{\xi-3}x, \ \ldots, \ H_{\xi-1}(x) = a_1 x$$

and for $n \geq 0$

$$H_{n+\xi}(x) = a_1 x H_{n+\xi-1}(x) + a_2 x H_{n+\xi-2}(x) + \cdots + a_{\xi-3}x H_{n+3}(x)$$
$$+ a_{\xi-2}x H_{n+2}(x) a_{\xi-1} H_{n+1}(x) + a_\xi H_n(x) \tag{2.1}$$

where $a_1, a_2, a_3 \ldots a_{\xi-1}, a_\xi$ are arbitrary algebraic integers and $\xi = 2, 3, 4, \ldots \infty$.

Note this notation $H_n^{[\xi]}(x)$ where ξ may be called the level of the Generalized Fibonacci Polynomials.

When $\xi = 2$, let us examine $H_n^{[2]}(x)$. On specialising the parameters, we obtain :

(i) Chebyshev Polynomials of second kind for

$$a_1 = 2, \qquad a_2 = -1,$$

(ii) Pell Polynomials for

$$a_1 = 2, \qquad a_2 = 1,$$

(iii) Fibonacci Polynomials $f_n(x)$ for

$$a_1 = 1, \qquad a_2 = 1,$$

When $\xi = 3$, with $H_n^{[3]}(x)$ we have the Jaiswal class of Polynomials.

When $\xi = 4$, with $H_n^{[4]}(x)$ we have the Harne-Parihar class of Polynomials.

When $\xi = 5$, with $\overset{[5]}{H_n}(x)$ we have

$$\overset{[5]}{H_0}(x) = 0, \overset{[5]}{H_1}(x) = 1,$$

$$\overset{[5]}{H_2}(x) = a_3 x, \overset{[5]}{H_3}(x) = a_2 x, \overset{[5]}{H_4}(x) = a_1 x$$

$$\overset{[5]}{H_5}(x) = \left(a_1^2 + a_2^2 + a_3^2\right)x^2 + a_4$$

$$\overset{[5]}{H_6}(x) = a_1\left(a_1^2 + a_2^2 + a_3^2\right)x^3 + \left(a_1 a_2 + a_2 a_3 + a_3 a_4\right)x^2 + a_5$$

$$\overset{[5]}{H_7}(x) = a_1 x\left[a_1\left(a_1^2 + a_2^2 + a_3^2\right)x^3 + \left(a_1 a_2 + a_2 a_3 + a_3 a_4\right)x^2 + a_5\right]$$
$$+ a_2 x\left[\left(a_1^2 + a_2^2 + a_3^2\right)x^2 + a_4\right] + a_1 a_3 x^2 + a_2 a_4 x + a_3 a_5 x$$
$$= a_1^2\left(a_1^2 + a_2^2 + a_3^2\right)x^4 + \left[a_2\left(a_1^2 + a_2^2 + a_3^2\right)\right.$$
$$\left. + a_1\left(a_1 a_2 + a_2 a_3 + a_3 a_4\right)\right]x^3$$
$$+ a_1 a_3 x^2 + \left(2a_2 a_4 + a_1 a_5 + a_3 a_5\right)x$$

On putting $x = 1$ and $\xi = 5$ in (2.1), we obtain another generalized Fibonacci sequence h_n where first few terms are :

$$\overset{[5]}{H_0} = 0, \ \overset{[5]}{H_1} = 1, \ \overset{[5]}{H_2} = a_3, \ \overset{[5]}{H_3} = a_2 \ \overset{[5]}{H_4} = a_1$$

$$\overset{[5]}{H_5} = a_1^2 + a_2^2 + a_3^2 + a_4$$

$$\overset{[5]}{H_6} = a_1\left(a_1^2 + a_2^2 + a_3^2\right) + \left(a_1 a_2 + a_2 a_3 + a_3 a_4\right) + a_5$$

$$\overset{[5]}{H_7} = a_1^2\left(a_1^2 + a_2^2 + a_3^2\right)$$
$$+ \left[a_2\left(a_1^2 + a_2^2 + a_3^2\right) + a_1\left(a_1 a_2 + a_2 a_3 + a_3 a_4\right)\right] + a_1 a_3$$
$$+ \left[2a_2 a_4 + a_1 a_5 + a_3 a_5\right]$$

We can continue generating infinite new Fibonacci sequences one for each value of $\xi = 5, 6, \ldots \infty$ and $x = 1$.

3. Generating Matrix for the Polynomials $H_n^{[\xi]}(x)$

Generating Matrix for the Polynomials $H_n^{[\xi]}(x)$, $[\xi = 2, 3, 4, \ldots \infty]$ is

$$R(x) = \begin{bmatrix} a_1 x & a_2 x & a_3 x & \ldots & a_{\xi-2} x & a_{\xi-1} & a_\xi \\ 1 & 0 & 0 & \ldots & 0 & 0 & 0 \\ 0 & 1 & 0 & \ldots & 0 & 0 & 0 \\ \hdotsfor{7} \\ 0 & 0 & 0 & \cdots & 1 & 0 & 0 \\ 0 & 0 & 0 & \cdots & 0 & 1 & 0 \end{bmatrix}$$

It can be easily proved by induction on n and ξ and that:
$R^n(x) =$

$$\begin{bmatrix} H_{n+\xi-2}(x) & {}^1G_{n+\xi-2}(x) & {}^2G_{n+\xi-2}(x) & \ldots & {}^{\xi-2}G_{n+\xi-2}(x) & {}^{\xi-1}G_{n+\xi-2}(x) \\ H_{n+\xi-3}(x) & {}^1G_{n+\xi-3}(x) & {}^2G_{n+\xi-3}(x) & \ldots & {}^{\xi-2}G_{n+\xi-3}(x) & {}^{\xi-1}G_{n+\xi-3}(x) \\ H_{n+\xi-4}(x) & {}^1G_{n+\xi-4}(x) & {}^2G_{n+\xi-4}(x) & \ldots & {}^{\xi-2}G_{n+\xi-4}(x) & {}^{\xi-1}G_{n+\xi-4}(x) \\ \hdotsfor{6} \\ H_n(x) & {}^1G_n(x) & {}^2G_n(x) & \ldots & {}^{\xi-2}G_n(x) & {}^{\xi-1}G_n(x) \\ H_{n-1}(x) & {}^1G_{n-1}(x) & {}^2G_{n-1}(x) & \ldots & {}^{\xi-2}G_{n-1}(x) & {}^{\xi-1}G_{n-1}(x) \end{bmatrix}$$

where

$$ {}^1G_n(x), \ {}^2G_n(x), \ {}^3G_n(x) \ldots {}^{\xi-2}G_n(x) \text{ and } {}^{\xi-1}G_n(x) $$

are the sequences of polynomials which obey the same relations but have different initial values.

$$ {}^1G_0(x) = 0, \ {}^1G_1(x) = 0, \ {}^1G_2(x) = 0, \ \ldots $$
$$ {}^1G_{\xi-3}(x) = 1, \ {}^1G_{\xi-2}(x) = 0, \ {}^1G_{\xi-1}(x) = a_2. $$
$$ {}^2G_0(x) = 0 \ {}^2G_1(x) = 0 \ {}^2G_2(x) = 0, \ \ldots $$
$$ {}^2G_{\xi-4}(x) = 1, \ {}^2G_{\xi-3}(x) = 0, \ {}^2G_{\xi-2}(x) = 0, \ {}^2G_{\xi-1}(x) = a_3. $$
$$ {}^9G_0(x) = 0, \ {}^9G_1(x) = 0, \ {}^9G_2(x) = 0, \ \ldots $$
$$ {}^9G_{\xi-5}(x) = 1, \ {}^9G_{\xi-4}(x) = 0, \ {}^9G_{\xi-3}(x) = 0, {}^9G_{\xi-2}(x) = 0, \ {}^9G_{\xi-1}(x) = a_4. $$
$$ \ldots\ldots\ldots\ldots\ldots\ldots\ldots\ldots\ldots\ldots\ldots\ldots $$

$$^{\xi-9}G_o(x) = 0, \ ^{\xi-9}G_1(x) = 1, \ ^{\xi-9}G_2(x) = 0, \ \dots$$

$$^{\xi-9}G_{\xi-2}(x) = 0, \ ^{\xi-9}G_{\xi-1}(x) = a_{\xi-2}.$$

$$^{\xi-2}G_0(x) = 1, \ ^{\xi-2}G_1(x) = 0, \ ^{\xi-2}G_2(x) = 0, \ \dots$$

$$^{\xi-2}G_{\xi-2}(x) = 0, \ ^{\xi-2}G_{\xi-1}(x) = a_{\xi-1}.$$

$$^{\xi-1}G_0(x) = 0, \ ^{\xi-1}G_1(x) = 0, \ ^{\xi-1}G_2(x) = 0, \ \dots$$

$$^{\xi-1}G_{\xi-2}(x) = 0, \ ^{\xi-1}G_{\xi-1}(x) = a_{\xi}. \quad [\xi = 2, 3, 4, \dots \infty]$$

We may succinctly state this as follows:

$$^{\tau}G_{\tau-i}(x) = 0 \ [\text{ for all } \ i \neq \tau, \ i \neq (\tau+1)]$$

and $^{\tau}G_{\tau-(\tau+1)}(x) = 1 \quad ^{\tau}G_{\tau}(x) = a_{\tau+1}$

$$[i = 1, 2, 3, \dots \tau.]$$
$$[\tau = 1, 2, 3, \dots (\xi-1).]$$
$$[\xi = 2, 3, 4, \dots \infty]$$

References

[1] Hoggatt, V. E. Jr. : *Fibonacci and Lucas Numbers*, Houshton Mifflin Company, Boston, 1969.

[2] Hoggatt, V. E. Jr., and D. A. Lind : Symbolic Substitution into Fibonacci Polynomials, *Fibonacci Quarterly*, Vol. 6, No.5, 1968, 55–74.

[3] Jaiswal D. V. : *A Study of Fibonacci sequences and polynomials*, Ph. D. Thesis, Mathematics, Univ. of Indore 1973. 169–182.

[4] Sanjay Harne and C. L. Parihar : Some Generalized Fibonacci Polynomials, *Journal Indian Academy of Mathematics.* Vol.18, No.2, 1996, 251–253.

A General Infinite Family of Congruence Relations

(Dedicated to my esteemed professor Shri. Faziludin Rowther, University College, Thiruvananthapuram.)

Abstract. A general infinite family of symmetric congruence relations are stated and proved. This generalizes the standard result
$$(-13)^{(n+1)} \equiv [(-13)^{(n)} + (-13)^{(n-1)}](\text{mod } 181).$$

Key Words:- Congruence Relations, Infinite.
AMS Subject Classification No:- 11A07.

1. A General Infinite Family of Congruence Relations

Theorem 1.1. *For "x" integer, $x \geq 2$, $n \geq 1$*

$$(-x)^{n+1} \equiv \left[(-x)^n + (-x)^{n-1}\right] \left[mod(x^2 + x - 1)\right]$$

$$(-x)^{n+3} \equiv \left[(-x)^{n+2} + (-x)^{n+1}(-x)^n + (-x)^{n-1}\right]$$
$$\left[mod(x^4 + x^3 - x^2 + x - 1)\right]$$

$$(-x)^{n+5} \equiv \left[(-x)^{n+4} + (-x)^{n+3} + \cdots + (-x)^n + (-x)^{n-1}\right]$$
$$\left[mod(x^6 + x^5 - x^4 + x^3 - x^2 + x - 1)\right]$$

....

$$(-x)^{n+2} \equiv \left[(-x)^{n+1} + (-x)^n + (-x)^{n-1}\right] \left[mod(-x^3 - x^2 + x - 1)\right]$$

$$(-x)^{n+4} \equiv \left[(-x)^{n+3} + (-x)^{n+2} + (-x)^{n+1} + (-x)^n + (-x)^{n-1}\right]$$
$$\left[mod(-x^5 - x^4 + x^3 - x^2 + x - 1)\right]$$

$$(-x)^{n+6} \equiv \left[(-x)^{n+5} + (-x)^{n+4} + \cdots + (-x)^n + (-x)^{n-1}\right]$$
$$\left[mod(-x^7 - x^6 + x^5 - x^4 + x^3 - x^2 + x - 1)\right]$$

....

Proof. Apply the principle of mathematical induction on the variables — First on "n" and then on "x" in each case. □

Generally

Theorem 1.2. *For "x" integer, $x \geq 2$, $n \geq 1$, $\xi \geq 1$ we have (for odd values $2\xi - 1$)*

$$(-x)^{(n+2\xi-1)} \equiv \left[(-x)^{(n+2\xi-2)} + (-x)^{(n+2\xi-3)} + \cdots + (-x)^n + (-x)^{n-1}\right]$$
$$\left[mod(-x^{2\xi} + x^{2\xi-1} - x^{2\xi-2} + x^{2\xi-3} - x^{2\xi-4} + \cdots + x - 1)\right]$$

Proof. Apply the principle of mathematical induction on the variables — First on "n" and then on "x" and then on "ξ". □

Note:

The standard result $(-13)^{(n+1)} \equiv \left[(-13)^{(n)} + (-13)^{(n-1)}\right](\text{mod } 181)$ is clearly a special case of Theorem 1.2 when $x = 13$ and $\xi = 1$ [**1**, p. 88].

Theorem 1.3. *For "x" integer, $x \geq 2$, $n \geq 1$, $\xi \geq 1$ we have (for even values 2ξ)*

$$(-x)^{(n+2\xi)} \equiv \left[(-x)^{(n+2\xi-1)} + (-x)^{(n+2\xi-2)} + \cdots + (-x)^n + (-x)^{n-1}\right]$$
$$\left[mod(-x^{2\xi+1} - x^{2\xi} + x^{2\xi-1} - x^{2\xi-2} + \cdots + x - 1)\right]$$

Proof. Apply the principle of mathematical induction on the variables — First on "n" and then on "x" and then on "ξ". □

References

[1] David M. Burton; *Elementary Number Theory*, Universal Book Stall, New Delhi, Second Edition. Reprint 1998.

Infinite Families of Base Systems and Properties of Numbers

(Dedicated to my esteemed teachers at University College, Thiruvananthapuram.)

Abstract. In this paper we deal with families of base systems and properties of numbers. Various standard results for base 10 are generalized appropriately for various base systems. The elaborate but elementary proof by the method of mathematical induction on two or three variables is omitted in many cases.

Key Words:- Base systems, Infinite Families.
AMS Subject Classification No:- 11A99.

1. Introduction

In this paper we deal with families of base systems and properties of numbers. For this we need a consistent notation for the Unit-Integers of any base system B for *All values of $B \geq 2$.*

Let $B \geq 2$ be any given base.

We define the Unit-Integers in this base system thus

$$\left(\boxed{0}, \boxed{1}, \boxed{2}, \boxed{3}, \boxed{4}, \boxed{5}, \cdots, \boxed{\text{B-3}}, \boxed{\text{B-2}}, \boxed{\text{B-1}}. \right)$$

For example let $B = 1729$.

We define the Unit-Integers in this base system thus

$$\left(\boxed{0}, \boxed{1}, \boxed{2}, \boxed{3}, \boxed{4}, \boxed{5}, \cdots, \boxed{1726}, \boxed{1727}, \boxed{1728}. \right)$$

Then the number

$$\boxed{134}\,\boxed{734}\,\boxed{894}\,\boxed{674}\,\boxed{583}\,\boxed{13}\,\boxed{10}_{[1729]}$$

$$= 10.(1729)^0 + 13.(1729)^1 + 583.(1729)^2$$
$$+ 674.(1729)^3 + 894.(1729)^4$$
$$+ 734.(1729)^5 + 134.(1729)^6$$

in the decimal system.

Theorem 1.1. *Let $N = a_m B^m + a_{m-1} B^{m-1} + \cdots + a_1 B + a_0$ be the representation of the positive integer N to the base B. $0 \leq a_k < B$, and let $R = a_0 - a_1 + a_2 - \cdots + (-1)^m a_m$.*
Then $(B+1)|N$ if and only if $(B+1)|R$.

Proof. Let $\mathfrak{P}(x) = \sum\limits_{k=0} a_k x^k$ be a polynomial with Integer Coefficients.

Since $B \equiv -1(\mathrm{mod}(B + 1))$, we get $\mathfrak{P}(B) \equiv \mathfrak{P}(-1)(\mathrm{mod}(B + 1))$. But $\mathfrak{P}(B) \equiv N$, whereas $\mathfrak{P}(-1) = a_0 - a_1 + a_2 - \cdots + (-1)^m a_m = R$, so that $N \equiv R(\mathrm{mod}(B + 1))$. This implies that either both N and R are divisible by $(B + 1)$ or neither is divisible by $(B + 1)$.
(For $B = 10$ we have the standard result.) [**1**, p. 93] $\square$

Theorem 1.2. *Let* $N = a_m 10^m + a_{m-1} 10^{m-1} + \cdots + a_1 10 + a_0$ *be the Decimal Expansion of the positive integer* N, $0 \le a_k < 10$, *and let* $J = a_0 + a_1 + a_2 + \cdots + a_m$.
Then $10^\xi - 1 | N$ *if and only if* $9\xi | J$. $[\xi = 1, 2, \cdots]$

Proof. For $\xi = 1$ we have the standard result [**1**, p. 93]. Prove by the application of the principle of mathematical induction on the variable "ξ". $\square$

Theorem 1.3. *Let* $N = a_m B^m + a_{m-1} B^{m-1} + \cdots + a_1 B + a_0$ *be the representation of the positive integer* N *to the base* B, $0 \le a_k < B$, *and let* $J = a_0 + a_1 + a_2 + \cdots + a_m$.
Then $(B^\xi - 1) | N$ *if and only if* $\xi(B - 1) | J$. $[\xi = 1, 2, \cdots]$

Proof. Prove by the application of the principle of mathematical induction on the variable "ξ". $\square$

Theorem 1.4. *Let* $F = \{B = 10\xi \; [\xi = 1, 2, \cdots]\}$ *be a family of base systems. Then for every integer "a" the unit digit of a^2 is*

$$\boxed{10\delta}, \boxed{10\delta + 1}, \boxed{10\delta + 4}, \boxed{10\delta + 5}, \boxed{10\delta + 6}$$

$$or \; \boxed{10\delta + 9}. \; [\delta = 0, 1, 2, \cdots (\xi - 1)]$$

in each member system in this family of base systems.

Proof. For $\xi = 1$ we have the standard result [**1**, p. 94]. Prove by the application of the principle of mathematical induction on the variable "ξ". $\square$

Theorem 1.5. *Let* $F = \{B \; [B = 2, 3, \cdots]\}$ *be a family of base systems. For every integer "a" the unit digit of a^3 can be any integer* $\boxed{x} < B$ *where B is the base of the system of notation.*

Proof. For $B = 10$ we have the standard result [**1**, p. 94]. Prove by the application of the principle of mathematical induction on the variable "B". $\square$

Theorem 1.6. *Let* $F = \{B \; [B = 2, 3, \cdots]\}$ *be a family of base systems. For every integer "a" the unit digit of a^τ [τ any odd number] can be any integer* $\boxed{x} < B$ *where B is the base of the system of notation.*

Proof. For $B = 10$ and $\tau = 3$ we have the standard result [**1**, p. 94]. Prove by the application of the principle of mathematical induction on the variable "B" and "τ". $\square$

Theorem 1.7. *Let $F = \{B = 10\xi \; [\xi = 1, 2, \cdots]\}$ be a family of base systems. Then for every integer "a" the unit digit of a^4 is*

$$\boxed{10\delta}, \boxed{10\delta + 1}, \boxed{10\delta + 5}, \; or \; \boxed{10\delta + 6}. \; [\delta = 0, 1, 2, \cdots (\xi - 1)]$$

in each member system in this family of base systems.

Proof. For $\xi = 1$ we have the standard result [**1**, p. 94]. Prove by the application of the principle of mathematical induction on the variable "ξ".   □

Theorem 1.8. *Let $F = \{B = 10\xi \; [\xi = 1, 2, \cdots]\}$ be a family of base systems. Then for every integer "a" the unit digit of $a^{4\tau}$ [$\tau \geq 1$] is*

$$\boxed{10\delta}, \boxed{10\delta + 1}, \boxed{10\delta + 5} \; or \; \boxed{10\delta + 6}. \; [\delta = 0, 1, 2, \cdots (\xi - 1)]$$

in each member system in this family of base systems.

Proof. Prove by the application of the principle of mathematical induction on the variables "ξ" and "τ".   □

Theorem 1.9. *Let $F = \{B = 10\xi \; [\xi = 1, 2, \cdots]\}$ be a family of base systems. Then the unit digit of a Triangular a Number is*

$$\boxed{10\delta}, \boxed{10\delta + 1}, \boxed{10\delta + 3}, \boxed{10\delta + 5}, \boxed{10\delta + 6}$$

$$or \; \boxed{10\delta + 8}. \; [\delta = 0, 1, 2, \cdots (\xi - 1)]$$

in each member system in this family of base systems.

Proof. For $\xi = 1$ we have the standard result [**1**, p. 94]. Prove by the application of the principle of mathematical induction on the variable "ξ".   □

Theorem 1.10. *Let $F = \{B = 2\xi \; [\xi = 1, 2, \cdots]\}$ be a family of base systems. Then any integer is divisible by $\boxed{2}$ if and only if its units digit is $\boxed{2\delta}$ [$\delta = 0, 1, 2, \cdots (\xi - 1)$] in each member system in this family of base systems.*

Proof. Prove by the application of the principle of mathematical induction on the variable "ξ". (For $\xi = 5$ we have the standard result [**1**, p. 94].)   □

Theorem 1.11. *Let $F = \{B = 3\xi + 1 \; [\xi = 1, 2, \cdots]\}$ be a family of base systems.*
Then any integer is divisible by $\boxed{\xi}$ if and only if the sum of its digits is divisible by $\boxed{\xi}$ in each member system in this family of base systems.

Proof. Prove by the application of the principle of mathematical induction on the variable "ξ". (For $\xi = 3$ we have the standard result [**1**, p. 94].)   □

Theorem 1.12. *Let $F = \{B = 3\xi + 1 \; [\xi = 3, 4, \cdots]\}$ be a family of base systems.*
Then any integer is divisible by the integer $\boxed{3}$ if and only if the sum of its digits is divisible by $\boxed{3}$ in each member system in this family of base systems.

Proof. Prove by the application of the principle of mathematical induction on the variable "ξ". (For $\xi = 3$ we have the standard result [**1**, p. 94].)   □

Theorem 1.13. *Let $F = \{B = 10\xi \; [\xi = 1, 2, \cdots]\}$ be a family of base systems. Then any integer of three or more digits is divisible by the integer $\boxed{4\delta}$ if and only if the number formed by its tens and units digits is divisible by $\boxed{4\delta}$ [$\delta = 1, 2, \cdots \xi$] in each member system in this family of base systems.*

Proof. For $\xi = 1$ we have the standard result [**1**, p. 94]. Prove by the application of the principle of mathematical induction on the variable "ξ". □

Theorem 1.14. *Let $F = \{B = 10\xi \; [\xi = 1, 2, \cdots]\}$ be a family of base systems. Then any integer of three or more digits is divisible by the integer $\boxed{4}$ if and only if the number formed by its tens and units digits is divisible by $\boxed{4}$ in each member system in this family of base systems.*

Proof. For $\xi = 1$ we have the standard result [**1**, p. 94]. Prove by the application of the principle of mathematical induction on the variable "ξ". □

Theorem 1.15. *Let $F = \{B = 2\xi \; [\xi = 1, 2, \cdots]\}$ be a family of base systems. Then any integer is divisible by $\boxed{\xi}$ if and only if its units digit is $\boxed{0}$ or $\boxed{\xi}$ in each member system in this family of base systems.*

Proof. Prove by the application of the principle of mathematical induction on the variable "ξ". (For $\xi = 5$ we have the standard result [**1**, p. 94].) □

Theorem 1.16. *Let $F = \{B \; [B = 2, 3, \cdots]\}$ be a family of base systems. Then the units digit of $\boxed{B\omega + 1}^{\tau}$ [$\omega = 0, 1, 2, \cdots$] [$\tau \geq 1$] is always, $\boxed{1}$ in each member system in this family of base systems.*

Proof. Prove by the application of the principle of mathematical induction on the variables "B" and "τ". □

Theorem 1.17. *Let $F = \{B = 2\xi \; [\xi = 2, 3, \cdots]\}$ be a family of base systems. Then the units digit of $\boxed{2\xi\omega + 2}^{\tau}$ [$\omega = 0, 1, 2, \cdots$] [$\tau \geq 1$] is always, $\boxed{2\delta}$. [$\delta = 1, 2, \cdots (\xi - 1)$] in each member system in this family of base systems.*

Proof. Prove by the application of the principle of mathematical induction on the variables "ξ" and "ω" and "τ". □

Theorem 1.18. *Let $F = \{B = 10\xi \; [\xi = 1, 2, \cdots]\}$ be a family of base systems. Then the units digit of $\boxed{10\xi\omega + 3}^{\tau}$ [$\omega = 0, 1, 2, \cdots$] [$\tau \geq 1$] is always, $\boxed{10\delta + 1}$, $\boxed{10\delta + 3}$, $\boxed{10\delta + 7}$ or $\boxed{10\delta + 9}$ [$\delta = 0, 1, 2, \cdots (\xi - 1)$] in each member system in this family of base systems.*

Proof. Prove by the application of the principle of mathematical induction on the variables "ξ" and "ω" and "τ". □

Theorem 1.19. *Let $F = \{B = 2(2\xi + 1) \; [\xi = 1, 2, \cdots]\}$ be a family of base systems. Then the units digit of $\boxed{2\xi}^{\tau}$ [$\tau \geq 1$] is always, $\boxed{2\xi}$ or $\boxed{2(\xi + 1)}$ in each member system in this family of base systems.*

Proof. Prove by the application of the principle of mathematical induction on the variables "ξ" and "τ". □

Theorem 1.20. *Let* $F = \{B = 2(2\xi + 1)\ [\xi = 1, 2, \cdots]\}$ *be a family of base systems.*

Then the units digit of $\boxed{2(2\xi + 1)\omega + 2\xi}^{\tau}$ *[$\omega = 0, 1, 2, \cdots$] [$\tau \geq 1$] is always,*

$\boxed{2\xi}$ *or* $\boxed{2(\xi + 1)}$ *in each member system in this family of base systems.*

Proof. Prove by the application of the principle of mathematical induction on the variables "ξ" and "τ". □

Theorem 1.21. *Let* $F = \{B = 4\xi\ [\xi = 1, 2, \cdots]\}$ *be a family of base systems. Then the units digit of* $\boxed{4\xi\omega + 2\xi}^{\tau}$ *[$\omega = 0, 1, 2, \cdots$] [$\tau \geq 1$] is always,* $\boxed{0}$ *or* $\boxed{2\xi}$ *in each member system in this family of base systems.*

Proof. Prove by the application of the principle of mathematical induction on the variables "ξ" and "ω" and "τ". □

Theorem 1.22. *Let* $F = \{B = 2(2\xi + 1)\ [\xi = 1, 2, \cdots]\}$ *be a family of base systems.*

Then the units digit of $\boxed{2(2\xi + 1)\omega + 2\xi + 1}^{\tau}$ *[$\omega = 0, 1, 2, \cdots$]*

[$\tau \geq 1$] $\boxed{2\xi + 1}$ *in each member system in this family of base systems.*

Proof. Prove by the application of the principle of mathematical induction on the variables "ξ" and "ω" and "τ". □

Theorem 1.23. *Let* $F = \{B = 2(2\xi + 1)\ [\xi = 1, 2, \cdots]\}$ *be a family of base systems.*

Then the units digit of $\boxed{2(\xi + 1)}^{\tau}$ *[$\tau \geq 1$] is always,* $\boxed{2(\xi + 1)}$ *in each member system in this family of base systems.*

Proof. Prove by the application of the principle of mathematical induction on the variables "ξ" and "τ". □

Theorem 1.24. *Let* $F = \{B = 2(2\xi + 1)\ [\xi = 1, 2, \cdots]\}$ *be a family of base systems.*

Then the units digit of $\boxed{2(2\xi + 1)\omega + 2(\xi + 1)}^{\tau}$ *[$\omega = 0, 1, 2, \cdots$]*

[$\tau \geq 1$] is always, $\boxed{2(\xi + 1)}$ *in each member system in this family of base systems.*

Proof. Prove by the application of the principle of mathematical induction on the variables "ξ" and "ω" and "τ". □

Theorem 1.25. *Let* $F = \{B = 10\xi\ [\xi = 1, 2, \cdots]\}$ *be a family of base systems. Then the units digit of* $\boxed{10\xi\omega + 7}^{\tau}$ *[$\omega = 0, 1, 2, \cdots$] [$\tau \geq 1$] is always,* $\boxed{10\delta + 1}$, $\boxed{10\delta + 3}$, $\boxed{10\delta + 7}$ *or* $\boxed{10\delta + 9}$ *[$\delta = 0, 1, 2, \cdots (\xi - 1)$] in each member system in this family of base systems.*

Proof. Prove by the application of the principle of mathematical induction on the variables "ξ" and "ω" and "τ". $\qquad\square$

Theorem 1.26. *Let $F = \{B = 10\xi \ [\xi = 1, 2, \cdots]\}$ be a family of base systems. Then the units digit of $\boxed{10\xi\omega + 8}^{\,\tau}$ [$\omega = 0, 1, 2, \cdots$] [$\tau \geq 1$] is always, $\boxed{10\delta + 2}$, $\boxed{10\delta + 4}$, $\boxed{10\delta + 6}$ or $\boxed{10\delta + 8}$. [$\delta = 0, 1, 2, \cdots (\xi - 1)$] in each member system in this family of base systems.*

Proof. Prove by the application of the principle of mathematical induction on the variables "ξ" and "ω" and "τ". $\qquad\square$

Theorem 1.27. *Let $F = \{B \ [B = 2, 3, \cdots]\}$ be a family of base systems. Then the units digit of $\boxed{B\omega + B - 1}^{\,\tau}$ [$\omega = 0, 1, 2, \cdots$] [$\tau \geq 1$] is always, $\boxed{1}$ or $\boxed{B - 1}$ in each member system in this family of base systems.*

Proof. Prove by the application of the principle of mathematical induction on the variables "B" and "ω" and "τ". $\qquad\square$

Theorem 1.28. *Let $F = \{B = 10\xi \ [\xi = 1, 2, \cdots]\}$ be a family of base systems. Then for any integer $a^2 - a + \boxed{7}$ the units digit is $\boxed{10\delta + 3}$, $\boxed{10\delta + 7}$ or $\boxed{10\delta + 9}$. [$\delta = 0, 1, 2, \cdots (\xi - 1)$] in each member system in this family of base systems.*

Proof. Prove by the application of the principle of mathematical induction on the variable "ξ". (For $\xi = 1$ we have the standard result [**1**, p. 94].) $\qquad\square$

Theorem 1.29. *Let $F = \{B = 10\xi \ [\xi = 1, 2, \cdots]\}$ be a family of base systems. Then for any integer $a^2 - a + \boxed{10\delta + 7}$ the units digit is $\boxed{10\delta + 3}$, $\boxed{10\delta + 7}$ or $\boxed{10\delta + 9}$. [$\delta = 0, 1, 2, \cdots (\xi - 1)$] in each member system in this family of base systems.*

Proof. Prove by the application of the principle of mathematical induction on the variable "ξ". $\qquad\square$

Theorem 1.30. *Let $F = \{B \ [B = 2, 3, \cdots]\}$ be a family of base systems. If T_N denotes the Nth triangular number, then $T_{N+2\lambda} \equiv T_N (mod \ \lambda)$; hence, T_N and T_{N+2B} must have the same last digit where B is any base in which the triangular numbers are notated.*

Proof. Prove by the application of the principle of mathematical induction on the variables "N" and "B". (For $B = 10$ we have the standard result [**1**, p. 95].) $\qquad\square$

Theorem 1.31. *Let $F = \{B = 2(2\xi + 1) \ [\xi = 1, 2, \cdots]\}$ be a family of base systems. 2^n divides an integer N if and only if 2^n divides the number made up of the last n digits of N in each member system in this family of base systems.*

Proof. Prove by the application of the principle of mathematical induction on the variable "ξ".
(Hint: $[2(2\xi + 1)]^k = 2^k(2\xi + 1)^k \equiv 0 \pmod{2^n}$ for $k \geq n$ and for all ξ.)
(For $\xi = 2$ we have the standard result [**1**, p. 95].) $\qquad\square$

Theorem 1.32. *Let* $F = \{B = 2(2\xi + 1) + (2\xi + 1)\omega \; [\xi = 1, 2, \cdots]$ $[\omega = 0, 1, 3, 5, \cdots]\}$ *be a family of base systems.*
$(\omega + 2)^n$ *divides an integer* N *if and only if* $(\omega + 2)^n$ *divides the number made up of the last* n *digits of* N *in each member system in this family of base systems.*

Proof. Prove by the application of the principle of mathematical induction on the variables "ξ" and "ω".
(Hint: $[2(2\xi + 1) + (2\xi + 1)\omega]^k = (\omega + 2)^k(2\xi + 1)^k \equiv 0 \pmod{(\omega + 2)^n}$ for $k \geq n$ and for all ξ and ω.)
(For $\xi = 2$ and $\omega = 0$ we have the standard result [**1**, p. 95].) $\qquad\square$

Theorem 1.33. *Let* $F = \{B \; [B = 2, 3, \cdots]\}$ *be a family of base systems.*
Then for any integer $N > 1$, $\exists$ *(there exists) a prime number with at least* N *of its digits equal to zero in each member system of this family of base systems.*

Proof. Prove by the application of the principle of mathematical induction on the variable "B".
(Hint: Consider the family of arithmetic progressions $B^{N+1}\lambda + 1$ for $\lambda = 1, 2, \cdots$ for each base B.)
(For $B = 10$ we have the standard result [**1**, p. 95].) $\qquad\square$

Theorem 1.34. *Let* $F = \{B = 3\xi + 1 \; [\xi = 3, 4, \cdots]\}$ *be a family of base systems.*
Let $N = a_m B^m + a_{m-1} B^{m-1} + \cdots + a_1 B + a_0$ *be the representation of the positive integer* N *to the base* B, $0 \leq a_k < B$.
Then $\boxed{2\xi + 1}$, $\boxed{3\xi + 2}$ *and* $\boxed{4\xi + 1}$ *all divide* N *if and only if* $\boxed{2\xi + 1}$, $\boxed{3\xi + 2}$ *and* $\boxed{4\xi + 1}$ *divide the integer*

$$M = \left[(3\xi + 1)^2 a_2 + (3\xi + 1)a_1 + a_0\right] - \left[(3\xi + 1)^2 a_5 + (3\xi + 1)a_4 + a_3\right]$$
$$+ \left[(3\xi + 1)^2 a_8 + (3\xi + 1)a_7 + a_6\right] - \cdots$$

in each member system in this family of base systems.

Proof. Prove by the application of the principle of mathematical induction on the variable "ξ".
Hint: If n is even, then

$$(3\xi + 1)^{3n} \equiv 1 \pmod{(3\xi + 1)^3 + 1};$$
$$(3\xi + 1)^{3n+1} \equiv (3\xi + 1) \pmod{(3\xi + 1)^3 + 1};$$
$$(3\xi + 1)^{3n+2} \equiv (3\xi + 1)^2 \pmod{(3\xi + 1)^3 + 1}.$$

If n is odd, then

$$(3\xi + 1)^{3n} \equiv -1(\mathrm{mod}\ (3\xi + 1)^3 + 1);$$

$$(3\xi + 1)^{3n+1} \equiv -(3\xi + 1)(\mathrm{mod}\ (3\xi + 1)^3 + 1);$$

$$(3\xi + 1)^{3n+2} \equiv -(3\xi + 1)^2(\mathrm{mod}\ (3\xi + 1)^3 + 1).$$

(For $\xi = 3$ we have the standard result [1, p. 95].) $\square$

Theorem 1.35. *Let* $F = \{B = 2(2\xi + 1)\ [\xi = 1, 2, \cdots]\}$ *be a family of base systems.*
Let $N = a_m B^m + a_{m-1} B^{m-1} + \cdots + a_1 B + a_0$ *be the representation of the positive integer* N *to the base* B, $0 \le a_k < B$.
Then $\boxed{2\xi + 1}$ *divides* N *if and only if* $\boxed{2\xi + 1}$ *divides the integer*

$$M = a_0 + \boxed{2\xi}a_1 + \boxed{2\xi}a_2 + \cdots + \boxed{2\xi}a_m.$$

in each member system in this family of base systems.

Proof. Prove by the application of the principle of mathematical induction on the variable "ξ".
(For $\xi = 2$ we have the standard result [1, p. 95].) $\square$

Theorem 1.36. *Let* $F = \{B = 2(2\xi + 1)\ [\xi = 1, 2, \cdots]\}$ *be a family of base systems.*
Then for any integer N, N *and* $N^{(2\xi+1)}$ *[$\xi = 1, 2, \cdots$] have the same unit's digit in each corresponding member system in this family of base systems.*

Proof. Prove by the application of the principle of mathematical induction on the variable "ξ".
(For $\xi = 2$ we have the standard result [1, p.119].) $\square$

Theorem 1.37. *Given an integer* N, *let* M *be the Reversido integer[1] formed by reversing the order of the digits of* N *in that base system* B *[$B \ge 2$].*
$(N - M)$ *is divisible by* $(B - 1)$ *in each of these base systems.*

Proof. Prove by the application of the principle of mathematical induction on the variable "B".
(For $B = 10$ we have the standard result [1, p.95].) $\square$

Theorem 1.38. *A palindrome is a number* N *that reads the same backwards as forwards in any base system* B *[$B \ge 2$]. i.e., When the Reversido integer* M *of an integer* N *is equal to itself it is palindrome.*
Any palindrome in any base system B *[$B \ge 2$] with an even number of digits is divisible by* $(B + 1)$ *in that base system.*

Proof. Prove by the application of the principle of mathematical induction on the variable "B".
(For $B = 10$ we have the standard result [1, p.95].) $\square$

---

[1]This terminology is new. I hope that it is acceptable!

Theorem 1.39. *Given a repunit R_n, represented to any base B [$B \geq 2$]*
  $(B-1)|R_n$ *if and only if* $(B-1)|n$.

Proof. Prove by the application of the principle of mathematical induction on the variable "B".
(For $B = 10$ we have the standard result [**1**, p.95].)   □

Theorem 1.40. *Given a repunit R_n, represented to any base B*
  $(B+1)|R_n$ *if and only if n is even.*

Proof. Prove by the application of the principle of mathematical induction on the variable "B".
(For $B = 10$ we have the standard result [**1**, p.95].)   □

Theorem 1.41. *Let I_n be an integer I repeated n times and called a repinteger I_n.*

Given a repunit R_n, or more generally a repinteger I_n represented to any base B
  $R_m|R_n$ *if and only if $m|n$. [for all $m > 1$ and $n > 1$]*
  $I_m|I_n$ *if and only if $m|n$. [for all $m > 1$ and $n > 1$]*

Proof. Trivial!   □

Theorem 1.42. *Let $\overset{\infty}{I}$ be repinteger-infinity i.e., an integer I repeated to infinity.*
i.e., $IIIIIIIIIIIIIIIIIIIIIIIIIIIIIII \cdots \infty$ [$I = 1, 2, \cdots \infty$]

   Every repinteger-infinity $\overset{\infty}{I}$ is divisible by all the factors f of the repeating integer I and it yields a repinteger-infinity $\overset{\infty}{J}$ with J having the same number of digits as I. When the number of digits of J is less than I then the vacant digit places are filled with zeroes.
   i.e.,

$$\frac{\overset{\infty}{I}}{f} = \overset{\infty}{J} \quad and \quad \frac{\overset{\infty}{I}}{\overset{\infty}{J}} = f.$$

For eg:

$$\overset{\infty}{128} = 128128128128128 \cdots \cdots \infty \ divided \ by \ 16 \ yields$$

$$008008008008008 \cdots \cdots \infty = \overset{\infty}{008}$$

i.e.,

$$\frac{\overset{\infty}{128}}{\overset{\infty}{008}} = 16.$$

Theorem 1.43.

a) A repinteger I_n [$I > 1$] and a repinteger-infinity $\overset{\infty}{I}$ [$I > 1$] represented in any base B are composite.

b) *Let $\overset{\infty}{[kx]}$ be the linearly rhythmic infinitely long integer [lrili] $\overset{\infty}{[kx]} = kx, 2kx,$*
   $3kx, \cdots \infty$ and let $[x]_n = 1kx, 2kx, 3kx, \cdots, nkx$ for all $k > 1$ and $x \geq 1$.
   For eg:

$$\overset{\infty}{[2.7]} = 714212835424956 \cdots \cdots \infty$$

$$\overset{\infty}{[5.7]} = 73570105140175 \cdots \cdots \infty$$

   for all $k > 1$ and $x \geq 1$, k is a factor of $\overset{\infty}{[kx]}$ and $[kx]_n$ and hence $\overset{\infty}{[kx]}$ and
   $[kx]_n$ represented in any base B are composite.

c) *Infinite possible functional zero-rhythms [fzor] {in fact any number of ze-*
   roes in any random fashion finite or infinite } may be inducted between the
   repeating integers in any repinteger-infinity [repi] $\overset{\infty}{I}$ [I > 1] and between
   the rhythmic integers in the sequences of any linearly rhythmic infinitely long
   integer [lrili] $\overset{\infty}{[kx]} = kx, 2kx, 3kx, \cdots \infty$ for all $k > 1$ and $x \geq 1$.
   All [fzor/lrili] and [fzor/repi] represented in any base B are composite.
   For eg:

   1. $3/\overset{\infty}{[kx]} = kx, 0_3 2kx, 0_3 3kx, 0_3 \cdots \infty$ for all $k > 1$ and $x \geq 1$

   2. $F/\overset{\infty}{[kx]} = kx, 0_{F(1)} 2kx, 0_{F(2)} 3kx, 0_{F(3)} \cdots \infty$ for all $k > 1$ and $x \geq 1$.
   [Here $F(\tau)$ is any function of a random variable τ. Many-valued func-
   tions and any generalized function may also be used to generate the zero
   rhythms.]

Notes:

Any infinitely long natural number N_∞ [limit integer] which is not a A repinteger-
infinity [repi] $\overset{\infty}{I}$ [I > 1] or a linearly rhythmic infinitely long integer [lrili]
$[\overset{\infty}{[kx]}]$ (for all $k > 1$ and $x \geq 1$) or [fzor/lrili] and [fzor/repi] is a limit prime
number.

Notes $\sim$

Any infinitely long natural number N_∞ [limit integer] which has no finite
factor other than 1 is a limit prime number.
i.e.,
Any infinitely long natural number N_∞ [limit integer] which is composite has at
least one finite factor other than 1.

References

[1] David M. Burton; *Elementary Number Theory*, Universal Book Stall, New Delhi, Second
    Edition. Reprint 1998.

Some General Properties of Arithmetic Functions and a Few Conjectures

(Dedicated to my esteemed teacher Shri Ananda Rao, SSKZM)

Abstract. Some General Properties of Well-Known Arithmetic Functions like $\sigma(n)$ and Euler's $\phi(n)$ function are presented. A few General Arithmetic Conjectures are also stated.

Key Words:- Arithmetic Functions, Generalizations.
AMS Subject Classification No:- 11A25.

1. Some Properties of Some Arithmetic Functions

Theorem 1.1. *For integers $k \geq 2$ and $\delta \geq 1$, if $\left(2^k - (2\delta - 1)\right)$ is Prime, then $n = 2^{k-1}\left(2^k - (2\delta - 1)\right)$ satisfies the equations*

$$\sigma(n) = 2n + (2\delta - 1) - 1 = 2(n + \delta - 1)$$

where $\sigma(n)$ denotes the Sum of the positive divisors of n.

Proof. Prove by the Application of the Principle of mathematical Induction on the variables "k" and "δ". (For $\delta = 1, 2$ we have the standard results.) [**1**, P.138]. □

Theorem 1.2. *For any integer $n \geq 1$ and any prime number p, if $\phi(n)$ is the Euler's ϕ function*

A) $\phi(pn) = p\phi(n)$ *if and only if $p|n$.*

B) $\phi(pn) = (p-1)\phi(n)$ *if and only if $p \nmid n$.*

and

C) $\phi\left(\prod_{i=1}^{k}(p_i)n\right) = \left\{\prod_{i=1}^{k}(p_i)\right\}\phi(n)$ *if and only if $p_i|n$ for all $i = 1, 2, \cdots k$.*

D) $\phi\left(\prod_{i=1}^{k}(p_i)n\right) = \left\{\prod_{i=1}^{k}(p_i - 1)\right\}\phi(n)$ *if and only if $p_i \nmid n$ for all $i = 1, 2, \cdots k$.*

Proof. Proofs follow from the elementary properties of the Euler's $\phi(n)$ function and elementary algebra. (For $p = 2, 3$, we have the standard results.) [**1**, p. 161] □

Theorem 1.3. *For any Prime Number p and Euler's $\phi(n)$ function,*

$$\phi(n) = \left\{ \frac{(p-1)}{2p} \right\} n$$

if and only if $n = 2^k p$ for some $k \geq 2$ ($k \geq 1$ for $p = 2$).

Proof. Proof follows from the elementary properties of the Euler's $\phi(n)$ function and elementary algebra. (For $p = 2$, we have the standard result.) [1, p. 161] $\square$

Theorem 1.4. *If $\phi(n)$ is the Euler's function, and if n is any positive integer and if the Prime Factorization of $n = p_1^{k_1}, p_2^{k_2} \cdots p_r^{k_r}$ for $\mu \geq 1$*

$$\sum_{i=1}^{\mu} \phi(n^i) = \phi(n) \left(\frac{(n^\mu - 1)}{(n-1)} \right)$$

$$\sum_{i=1}^{\mu} \phi(n^i) = \left(\prod_{t=1}^{r} p_t^{k_t} \right) \left(\prod_{t=1}^{r} \left(1 - \frac{1}{p_t} \right) \right) \left(\frac{\left(\prod_{t=1}^{r} p_t^{k_t} \right)^\mu - 1}{\left(\prod_{t=1}^{r} p_t^{k_t} \right) - 1} \right).$$

Proof. Proof follows from the elementary properties of the Euler's $\phi(n)$ function and elementary algebra. $\square$

Theorem 1.5. *Let integer $n \geq \phi$. We could define the Polynomial Sum of the positive divisors of n with unit coefficients.*
   Let

$$\sigma_{\sum_{i \sim s}}(n) = \sum_{d|n} \sum_{i=0}^{s} d^i$$

(Here $\tau(n)$ denotes the number of positive divisors of n and $\sigma(n)$ denotes the sum of the positive divisors as usual.)
then

1) $\sigma_{\sum_{0 \sim 0}}(n) = \tau(n)$ and $\sigma_{\sum_{0 \sim 1}}(n) = \tau(n) + \sigma(n)$.

2) $\sigma_{\sum_{0 \sim s}}(n)$ is a multiplicative function.

3) If $n = p_1^{k_1}, p_2^{k_2}, p_3^{k_3} \cdots p_r^{k_r}$ is the prime factorization of n

$$\sigma_{\sum_{0 \sim s}}(n) = \tau(n) + \sum_{i=1}^{s} \left(\frac{p_1^{i(k_1+1)} - 1}{p_1^i - 1} \right) \left(\frac{p_2^{i(k_2+1)} - 1}{p_2^i - 1} \right) \cdots \left(\frac{p_r^{i(k_r+1)} - 1}{p_r^i - 1} \right).$$

Proof. Standard Algebra. $\square$

Theorem 1.6. *Let $n \geq \phi$, $a_i \geq \phi$ for $i = 1, 2, \cdots s$ be integers. We could define the Polynomial Sum of the positive divisors with integer coefficients.*
   Let

$$\sigma_{\sum a_{i \sim s}}(n) = \sum_{d|n} \sum_{i=0}^{s} a_i d^i$$

(Here $\tau(n)$ denotes the number of positive divisors of n and $\sigma(n)$ denotes the sum of the positive divisors as usual.)
then

1) $\sigma_{\sum a_{0\sim 0}}(n) = a_0\tau(n)$ *and* $\sigma_{\sum a_{0\sim 1}}(n) = a_0\tau(n) + a_i\sigma(n)$.

2) $\sigma_{\sum a_{i\sim s}}(n)$ *is a multiplicative function.*

3) If $n = p_1^{k_1}, p_2^{k_2}, p_3^{k_3} \cdots p_r^{k_r}$ *is the prime factorization of* n

$$\sigma_{\sum a_{i\sim s}}(n) = a_0\tau(n) + \sum_{i=1}^{s} a_i$$
$$\left\{ \left(\frac{p_1^{i(k_1+1)} - 1}{p_1^i - 1} \right) \left(\frac{p_2^{i(k_2+1)} - 1}{p_2^i - 1} \right) \cdots \left(\frac{p_r^{i(k_r+1)} - 1}{p_r^i - 1} \right) \right\}.$$

Proof. Standard Algebra. $\qquad\square$

Special Cases

1) When all a_i $[i = 0 \cdots (s-1)]$ are zero and $a_s = 1$ we have the case

$$\sigma_{\sum a_{0\sim 0}}(n) = a_0\tau(n) = \phi$$
$$\sigma_{\sum a_{0\sim 1}}(n) = a_0\tau(n) + a_1\sigma(n) = \phi$$
$$\sigma_{\sum a_{i\sim s}}(n) = \left(\frac{p_1^{s(k_1+1)} - 1}{p_1^s - 1} \right) \left(\frac{p_2^{s(k_2+1)} - 1}{p_2^s - 1} \right) \cdots \left(\frac{p_r^{s(k_r+1)} - 1}{p_r^s - 1} \right).$$

2) When all $a_i(i = 2 \cdots (s-1))$ are zero and $a_0, a_1, a_s = 1$ we have the case

$$\sigma_{\sum a_{0\sim 0}}(n) = \tau(n)$$
$$\sigma_{\sum a_{0\sim 1}}(n) = \tau(n) + \sigma(n)$$
$$\sigma_{\sum a_{i\sim s}}(n) = \tau(n) + \sigma(n) +$$
$$\left(\frac{p_1^{s(k_1+1)} - 1}{p_1^s - 1} \right) \left(\frac{p_2^{s(k_2+1)} - 1}{p_2^s - 1} \right) \cdots \left(\frac{p_r^{s(k_r+1)} - 1}{p_r^s - 1} \right).$$

These are extensions of standard results. [**1**, p. 139]

2. Some General Conjectures

Conjecture 2.1. Let $F = \{B = 2\xi \ [\xi = 1, 2, \cdots]\}$ be Family of Base Systems. Let $R_B = \{R_i\}_B \ [i = 1, 2, \cdots \infty]$ be the set of All Repunits to any Base in Each Member System in this Family of Base Systems.

For each value of $\xi \geq 1$, the Set $R_B = \{R_i\}_B$ contains infinite Prime Numbers.

Note: For $B = 10$ there are very few Repunits–Primes detected so far.
But I persume we have allways just begun to count, however big the number we are dealing with.

So the Conjecture clearly demands a contemptuous disproof or invites a beautiful inductive proof.

Conjecture 2.2. Let

$$\overset{1}{\underset{[a_1,b_1]}{D}} = a_1 + (n-1)b_1 = \left\{ \overset{1}{\underset{1}{s}}, \overset{1}{\underset{2}{s}}, \cdots, \overset{1}{\underset{n}{s}}, \cdots \infty \right\}$$

$$\overset{2}{\underset{[a_2,b_2]}{D}} = a_2 + (n-1)b_2 = \left\{ \overset{2}{\underset{1}{s}}, \overset{2}{\underset{2}{s}}, \cdots, \overset{2}{\underset{n}{s}}, \cdots \infty \right\}$$

$$\overset{3}{\underset{[a_3,b_3]}{D}} = a_3 + (n-1)b_3 = \left\{ \overset{3}{\underset{1}{s}}, \overset{3}{\underset{2}{s}}, \cdots, \overset{3}{\underset{n}{s}}, \cdots \infty \right\}$$

$$\cdots\cdots\cdots\cdots\cdots\cdots\cdots\cdots$$

$$\overset{i}{\underset{[a_i,b_i]}{D}} = a_i + (n-1)b_i = \left\{ \overset{i}{\underset{1}{s}}, \overset{i}{\underset{2}{s}}, \cdots, \overset{i}{\underset{n}{s}}, \cdots \infty \right\} \quad [i \geq 2]$$

be a family $D = \left\{ \overset{i}{\underset{[a_i,b_i]}{D}} \right\}$ of Dirichlet Arithmetic Progressions when a_i and b_i are relatively prime for each value of i and hence and such that there are infinite prime numbers in the sequence $\left\{ \overset{i}{\underset{1}{s}}, \overset{i}{\underset{2}{s}}, \cdots, \overset{i}{\underset{n}{s}}, \cdots \infty \right\}$ for each value of $i [i \geq 2]$.

If $\overset{i}{\underset{p}{s}}$ is prime for each each value of i for some p.

Then we would call $\left\{ \overset{1}{\underset{p}{s}}, \overset{2}{\underset{p}{s}}, \overset{3}{\underset{p}{s}}, \cdots, \overset{i}{\underset{p}{s}} \right\}$ parallely prime.

If

$$D = \left\{ \overset{i}{\underset{[a_i,b_i]}{D}} \right\} = a_i + (n-1)b_i = \left\{ \overset{i}{\underset{1}{s}}, \overset{i}{\underset{2}{s}}, \cdots, \overset{i}{\underset{n}{s}}, \cdots \infty \right\} \quad [i \geq 2]$$

is parallely prime for some $p \geq 1$. Then $D = \left\{ \overset{i}{\underset{[a_i,b_i]}{D}} \right\}$ is parallely prime for infinite values of n.

Conjecture 2.3. Let $D = \left\{ \overset{i}{\underset{[a_i,b_i]}{D}} \right\}$ be a family of Dirichlet Arithmetic Progressions when a_i and b_i are relatively prime for each value of i and hence and such that there are infinite prime numbers in the sequence $\left\{ \overset{i}{\underset{1}{s}}, \overset{i}{\underset{2}{s}}, \cdots, \overset{i}{\underset{n}{s}}, \cdots \infty \right\}$ for each value of $i [i \geq 2]$.

If $\overset{i}{\underset{p}{s}}$ is prime for each each value of i for some p.

Then we would call $\left\{ \overset{1}{\underset{p}{s}}, \overset{2}{\underset{p}{s}}, \overset{3}{\underset{p}{s}}, \cdots, \overset{i}{\underset{p}{s}} \right\}$ parallely prime.

$$D = \left\{ \underset{[a_i,b_i]}{\overset{i}{D}} \right\} = a_i + (n-1)b_i = \left\{ \overset{i}{\underset{1}{s}}, \overset{i}{\underset{2}{s}}, \cdots, \overset{i}{\underset{n}{s}}, \cdots \infty \right\} \quad [i \geq 2]$$

is parallely prime for infinite values of n.

Note:

Wilson's Theorem implies that $\exists$ infinite composite numbers of the form $(n! + 1)$. On the other hand, it is an open question whether $(n! + 1)$ is prime for infinitely many values of n. [**1**, p. 125]

Conjecture 2.4.
$\exists$ infinite composite numbers of the form $[n! + (2\xi - 1)], \forall \, \xi \geq 1$.

Conjecture 2.5.
$\exists$ infinite prime numbers of the form $[n! + (2\xi - 1)], \forall \, \xi \geq 1$.

Conjecture 2.6.
$\exists$ infinite composite numbers of the form $[n! - (2\xi - 1)], \forall \, \xi \geq 1$.

Conjecture 2.7.
$\exists$ infinite prime numbers of the form $[n! - (2\xi - 1)], \forall \, \xi \geq 1$.

References

[1] David M. Burton; *Elementary Number Theory*, Universal Book Stall, New Delhi, Second Edition. Reprint 1998.

Some General Divisibility Statements

(Dedicated to my esteemed teachers at SSKZM)

Abstract. A few General Divisibility Statements which can be proved by Mathematical Induction are stated. These generalize existing results.

Key Words:- Divisibility Statements, Infinite Generalizations.
AMS Subject Classification No:- 11A99.

A General Divisibility Statement

Theorem 0.1.

$$\prod_{\xi=0}^{n} \left[(n+1)^2 - \xi^2\right] \;\Bigg|\; \prod_{\xi=0}^{n} \left[a^2 - \xi^2\right]$$

or equivalently

$$\{[(2n+1)!] \; (n+1)\} \;\Bigg|\; \prod_{\xi=0}^{n} \left[a^2 - \xi^2\right]$$

for all integers a and n such that $n \geq 0$.

Proof. Prove by the Application of the Principle of Mathematical Induction on the variable "n" for all integers "a".
For $n = 1$ and $n = 2$ we have the standard results [1, p.34]

$$12 \;\Bigg|\; \prod_{\xi=0}^{1} \left[a^2 - \xi^2\right] \quad \text{and} \quad 360 \;\Bigg|\; \prod_{\xi=0}^{2} \left[a^2 - \xi^2\right]. \qquad \square$$

Theorem 0.2.

$$\prod_{\xi=0}^{n} \left[(n+1)^2 - \xi^2\right] \;\Bigg|\; \prod_{\xi=0}^{n} \left[a^{2\tau} - \xi^2\right]$$

or equivalently

$$\{[(2n+1)!] \; (n+1)\} \;\Bigg|\; \prod_{\xi=0}^{n} \left[a^{2\tau} - \xi^2\right]$$

for all integers a, n and τ such that $n \geq 0$, $\tau \geq 1$.

Proof. Substitute $a = a^\tau$ in Theorem 0.1. $\qquad \square$

Theorem 0.3. *Another General Divisibility Statement* $\sim$

$$2^{\xi} \;\Big|\; \left\{ 5^{\left[2^{\xi-2}\right]n} + \left[2^{\xi} - 1\right] \right\}$$

for all integers a and n such that $\xi \geq 2 \; n \geq 1$.

   The standard results $8 \big| (5^{2n} + 7)$ now becomes the special case of theorem 0.3 when $\xi = 3$ [1, p. 32].

Proof. Prove by Mathematical Induction on ξ.

   [Hint $\sim$

$$5^{\left[2^{(\xi-2)}\right][k+1]} + \left[2^{\xi} - 1\right] = 5^{\left[2^{(\xi-2)}\right]} \left[5^{\left[2^{(\xi-2)}k\right]} + \left[2^{\xi} - 1\right]\right]$$
$$+ \left[\left[2^{\xi} - 1\right] - \left\{5^{\left[2^{(\xi-2)}\right]} \cdot \left[2^{\xi} - 1\right]\right\}\right]$$

for all $\xi \geq 2$.] $\qquad\qquad\qquad\qquad\qquad\qquad\qquad\qquad \square$

References

[1] David M. Burton; *Elementary Number Theory*, Universal Book Stall, New Delhi, Second Edition. Reprint 1998.

Some General Results in Elementary Number Theory

(Dedicated to all my teachers in all the schools I studied)

Abstract. Some results in Elementary Number Theory are generalized. All the proofs use the principle of mathematical induction. Induction on two or even three distinct variables are involved in each case. In order to save space the details of the proofs are omitted in many cases.

Key Words:- Mathematical Induction, Infinity.
AMS Subject Classification No:- 11A99.

1. Some Simple Results

Theorem 1.1.

1) $0 + 1 + 2 + 3 + \cdots + (1n - 1)^2 = n^2 - 1 \left(\frac{n(n+1)}{2} \right)$ *(Standard result 1,2)*

2) $1 + 3 + 5 + 7 + \cdots + (2n - 1)^2 = n^2 + 0 \left(\frac{n(n+1)}{2} \right)$ *(Standard result 1,2)*

Trivially generalizing we can prove by Mathematical Induction

3) $2 + 5 + 7 + 9 + \cdots + (3n - 1)^2 = n^2 + 1 \left(\frac{n(n+1)}{2} \right)$

4) $3 + 7 + 11 + 15 + \cdots + (4n - 1)^2 = n^2 + 2 \left(\frac{n(n+1)}{2} \right)$

....

ξ) $(\xi - 1) + (2\xi - 1) + \cdots + (\xi n - 1)^2 = n^2 + (\xi - 2) \left(\frac{n(n+1)}{2} \right)$ *(ξ +ve integer)*

$$\sum_{j=1}^{n} (\xi j - 1)^2 = n^2 + (\xi - 2) \left(\frac{n(n+1)}{2} \right) \quad (\xi = 1, 2, 3, \cdots \infty).$$

Re-writing in the notation for the super-sums (Sum level indicated on top) we have

$$\left[\overset{[1]}{(\xi n - 1)^2} \right] = (\xi n - 1)^2$$

$$S_n \left[\overset{[1]}{(\xi n - 1)^2} \right] = n^2 + (\xi - 2) \left(\frac{n(n+1)}{2} \right) = \left[\overset{[2]}{(\xi n - 1)^2} \right] \tag{1.1}$$

The sum of the above sequence may be curiously written as

$$S_n \left[(\xi n \overset{[2]}{-} 1)^2 \right] = (\xi - 1) + (3\xi - 2)(n-1) + ((3\xi - 1) + \xi(n - \xi))(n - 2)$$

$$= \left[(\xi n \overset{[3]}{-} 1)^2 \right]$$

$$S_n \left[(\xi n \overset{[3]}{-} 1)^2 \right] = (\xi - 1)n + (3\xi - 2)\left(\frac{(n-1)(n)}{2!} \right)$$

$$+ \left((3\xi - 1) + \frac{\xi(n-\xi)(n-\xi+1)}{2!} \right) \left(\frac{(n-2)(n-1)}{2!} \right)$$

$$= \left[(\xi n \overset{[4]}{-} 1)^2 \right]$$

$$S_n \left[(\xi n \overset{[4]}{-} 1)^2 \right] = (\xi - 1)\left(\frac{n(n+1)}{2!} \right) + (3\xi - 2)\left(\frac{(n-1)n(n+1)}{3!} \right)$$

$$+ \left((3\xi - 1) + \frac{\xi(n-\xi)(n-\xi+1)(n-\xi+2)}{3!} \right)$$

$$\left(\frac{(n-2)(n-1)n}{3!} \right)$$

$$= \left[(\xi n \overset{[5]}{-} 1)^2 \right]. \tag{1.2}$$

....

$$S_n \left[(\xi n \overset{[\alpha]}{-} 1)^2 \right] = (\xi - 1)\left(\frac{n(n+1)\cdots(n+\alpha-3)}{(\alpha-2)!} \right)$$

$$+ (3\xi - 2)\left(\frac{(n-1)n\cdots(n+\alpha-3)}{(\alpha-1)!} \right)$$

$$+ \left((3\xi - 1) + \frac{\xi(n-\xi)(n-\xi+1)\cdots(n-\xi+\alpha-2)}{(\alpha-1)!} \right)$$

$$\left(\frac{(n-2)(n-1)n\cdots(n+\alpha-4)}{(\alpha-1)!} \right) = \left[(\xi n \overset{[\alpha+1]}{-} 1)^2 \right].$$
$$\tag{1.3}$$

(ξ +ve integer: $(n-i)$ is operative only for $(n-i) \geq 0$)

Proof. Apply the principle of mathematical induction on the variables "n", "ξ" and finally "α". $\qquad\square$

Theorem 1.2.

$$\left[(\xi n - \overset{[1]}{(\xi - 1)})^2\right] = 1 + [\xi^2 + 2\xi](n-1) + [2\xi^2]\frac{(n-2)(n-1)}{2!}$$

$$S_n\left[(\xi n - \overset{[1]}{(\xi - 1)})^2\right] = n + [\xi^2 + 2\xi]\frac{(n-1)(n)}{2!} + [2\xi^2]\frac{(n-2)(n-1)(n)}{3!}$$

$$= \left[(\xi n - \overset{[2]}{(\xi - 1)})^2\right]$$

$$S_n\left[(\xi n - \overset{[2]}{(\xi - 1)})^2\right] = \frac{n(n+1)}{2!} + [\xi^2 + 2\xi]\frac{(n-1)(n)(n+1)}{3!}$$

$$+ [2\xi^2]\frac{(n-2)(n-1)(n)(n+1)}{4!}$$

$$= \left[(\xi n - \overset{[3]}{(\xi - 1)})^2\right]$$

$$S_n\left[(\xi n - \overset{[3]}{(\xi - 1)})^2\right] = \frac{n(n+1)(n+2)}{3!} + [\xi^2 + 2\xi]\frac{(n-1)(n)(n+1)(n+2)}{4!}$$

$$+ [2\xi^2]\frac{(n-2)(n-1)(n)(n+1)(n+2)}{5!}$$

$$= \left[(\xi n - \overset{[4]}{(\xi - 1)})^2\right]$$

$$\cdots\cdots\cdots\cdots\cdots\cdots\cdots\cdots\cdots$$

$$S_n\left[(\xi n - \overset{[\alpha]}{(\xi - 1)})^2\right] = \frac{n(n+1)\cdots(n+\alpha-1)}{\alpha!}$$

$$+ [\xi^2 + 2\xi]\frac{(n-1)(n)\cdots(n+\alpha-1)}{(\alpha+1)!}$$

$$+ [2\xi^2]\frac{(n-2)(n-1)\cdots(n+\alpha-1)}{(\alpha+2)!}$$

$$= \left[(\xi n - \overset{[\alpha+1]}{(\xi - 1)})^2\right]$$

(ξ +ve integer: $(n-i)$ is operative only for $(n-i) \geq 0$).

Proof. Apply the principle of mathematical induction on the variables "n", "ξ" and finally "α".   □

Theorem 1.3.

$$\left[(\xi n \overset{[1]}{-} \Delta)^2\right] = [\xi - \Delta]^2 + [3\xi^2 - 2\xi\Delta](n-1) + [2\xi^2]\frac{(n-2)(n-1)}{2!}$$

$$S_n\left[(\xi n \overset{[1]}{-} \Delta)^2\right] = [\xi - \Delta]^2(n) + [3\xi^2 - 2\xi\Delta]\frac{(n-1)(n)}{2!}$$

$$+ [2\xi^2]\frac{(n-2)(n-1)(n)}{3!} = \left[(\xi n \overset{[2]}{-} \Delta)^2\right]$$

$$S_n\left[(\xi n \overset{[2]}{-} \Delta)^2\right] = [\xi - \Delta]^2\frac{n(n+1)}{2!} + [3\xi^2 - 2\xi\Delta]\frac{(n-1)(n)(n+1)}{3!}$$

$$+ [2\xi^2]\frac{(n-2)(n-1)(n)(n+1)}{4!}$$

$$= \left[(\xi n \overset{[3]}{-} \Delta)^2\right]$$

$$S_n\left[(\xi n \overset{[3]}{-} \Delta)^2\right] = [\xi - \Delta]^2\frac{n(n+1)(n+2)}{3!}$$

$$+ [3\xi^2 - 2\xi\Delta]\frac{(n-1)(n)(n+1)(n+2)}{4!}$$

$$+ [2\xi^2]\frac{(n-2)(n-1)(n)(n+1)(n+2)}{5!}$$

$$= \left[(\xi n \overset{[4]}{-} \Delta)^2\right]$$

....

$$S_n\left[(\xi n \overset{[\alpha]}{-} \Delta)^2\right] = [\xi - \Delta]^2\frac{n(n+1)\cdots(n+\alpha-1)}{\alpha!}$$

$$+ [3\xi^2 - 2\xi\Delta]\frac{(n-1)(n)\cdots(n+\alpha-1)}{(\alpha+1)!}$$

$$+ [2\xi^2]\frac{(n-2)(n-1)\cdots(n+\alpha-1)}{(\alpha+2)!}$$

$$= \left[(\xi n \overset{[\alpha+1]}{-} \Delta)^2\right]$$

("ξ" +ve integer: "Δ" any integer: $(n-i)$ is operative only for $(n-i) \geq 0$).

Proof. Apply the principle of mathematical induction on the variables "n", "ξ" and finally "α". □

2. Some Simple Generalizations

Theorem 2.1. *For n any +ve even integer we have*

$$1.\,(n) + 2.\,(n-1) + 3.\,(n-2) + \cdots + \left(\frac{n}{2}\right)\left(\frac{n}{2}+1\right)$$

$$= \frac{n(n+1)(n+2)}{12} = \frac{1}{2}\left[\sum_{\lambda=1}^{n}\sum\lambda\right].$$

Proof. Apply the principle of mathematical induction on the variable "n". □

Theorem 2.2. *For n any +ve even integer and ∂ any factor of n we have*

$$(\partial \cdot n) + 2\partial \cdot (n-\partial) + 3\partial \cdot (n-2\partial) + \cdots + \frac{n}{2}\left(\frac{n}{2}+\partial\right) = \frac{n^3 + 3n^2\partial + 2n\partial^2}{12\partial}.$$

Proof. Apply the principle of mathematical induction on the variable "n". □

Theorem 2.3. *For all integers $n \geq 2\lambda$, $n! > n^\lambda$ ($\lambda = 1, 2, 3, \cdots \infty$).*

Proof. Apply the principle of mathematical induction on the variables "n" and "λ". (For $\lambda = 2$ and $\lambda = 3$ we have the standard results.) [1, p.9]. □

Theorem 2.4. *For all $\lambda \geq 2$ and $n > 1$ ("λ" and "n" integers)*

$$\frac{1}{1^\lambda} + \frac{1}{2^\lambda} + \frac{1}{3^\lambda} + \cdots + \frac{1}{n^\lambda} < 2 - \left(\frac{1}{n} + \frac{1}{n^2} + \cdots + \frac{1}{n^{(\lambda-1)}}\right).$$

Proof. Apply the principle of mathematical induction on the variables "n" and "λ". (For $\lambda = 2$ we have the standard results.) [1, p.10]. □

Theorem 2.5.

$$a^2 + (a+1)^2 + (a+1)^2 + (a+2)^2 + (a+2)^2 + (a+2)^2 + \cdots$$
$$(a + (n-1))^2 + (a + (n-1))^2 \cdots n \text{ times}$$
$$= a^2\left\{\frac{(n)(n+1)}{2}\right\} + 2a\left\{\frac{(n-1)(n)(n+1)}{3}\right\} +$$
$$\left\{\frac{(n-1)(n)}{2}\right\}^2 + \left\{\frac{(n-1)(n)(2n-1)}{6}\right\}$$

for all algebraic values of "a" and $n \geq 1$, n integer.

Proof. Apply the principle of mathematical induction on the variable "n". □

Theorem 2.6. *For $n > 3$ integers n, $(n + 2^\lambda)$, $(n + 2^{\lambda+1})$ ($\lambda = 1, 2, 3, \cdots \infty$) cannot all be primes.*

Proof. Apply the principle of mathematical induction on the variables "n" and "λ". (For $\lambda = 1$ we have the standard result.) [1, p.75]. □

Theorem 2.7. *For $n > 3$ integers*

$$\left(n + \sum_{i=1}^{t} 2^{(\lambda+i)}\right), \left(n + \sum_{i=0}^{t} 2^{(\lambda+i)}\right), (n + 2^{(\lambda+t+1)}),$$

[$\lambda = 1, 2, 3, \cdots \infty$], [$t = 1, 2, 3, \cdots \infty$] cannot all be primes.

Proof. Apply Theorem 2.6 repeatedly. □

Theorem 2.8. *For $n > 3$, let n, $(n+2)$, $(n+4)$, $\cdots$, $(n+2\delta)$ [$\delta = 2, 3, \cdots \infty$], be a Family of sequences of integers $\{S_\delta\}$.*

Let $P = \{{}^{\delta}P_i\} = \{{}^{\delta}P_1, {}^{\delta}P_2, \cdots, {}^{\delta}P_t\}$ be the set of all odd prime numbers such that ${}^{\delta}P_i \leq \delta + 1$ for each i for each S_δ.

Then each of the sequences of $\{S_\delta\}$ [$\delta = 2, 3, \cdots \infty$], contains utmost

$$(\delta + 1)\left[1 - \sum_{i=1}^{t} \frac{1}{{}^{\delta}P_i} + \sum_{\substack{i,j=1 \\ i \neq j}}^{t} \frac{1}{{}^{\delta}P_i \, {}^{\delta}P_j}\right] \quad \text{prime numbers.}$$

Proof. Apply the principle of mathematical induction on the variables "n" and "δ". □

Theorem 2.9. *For $n > 3$, let n, $(n + 2^\xi)$, $\cdots (n + 2^\xi\delta)$ [$\xi = 1, 2, 3, \cdots \infty$], [$\delta = 2, 3, \cdots \infty$], be a Families of sequences of integers $\{{}^{\xi}S_\delta\}$ for each ξ.*

Let $P = \{{}^{\delta}P_i\} = \{{}^{\delta}P_1, {}^{\delta}P_2, \cdots, {}^{\delta}P_t\}$ be the set of all odd prime numbers such that ${}^{\delta}P_i \leq \delta + 1$ for each i for each S_δ.

Then each of the sequences of $\{{}^{\xi}S_\delta\}$ [$\xi = 1, 2, 3, \cdots \infty$], [$\delta = 2, 3, \cdots \infty$] contains utmost

$$(\delta + 1)\left[1 - \sum_{i=1}^{t} \frac{1}{{}^{\delta}P_i} + \sum_{\substack{i,j=1 \\ i \neq j}}^{t} \frac{1}{{}^{\delta}P_i \, {}^{\delta}P_j}\right] \quad \text{prime numbers.}$$

Proof. Apply the principle of mathematical induction on the variables "n" and "δ" and ξ. □

Theorem 2.10. *For every integer $\delta \geq 2$ and for each integer $n \geq 1$*

$$(\delta!) \left| \frac{(\delta n)!}{(n!)^\delta} \right. .$$

Proof. Apply the principle of mathematical induction on the variables "n" and "δ". (For $\delta = 2$ we have the standard result.) [**1**, p.151]. □

Theorem 2.11. *For every integer $\delta_i \geq 2$ [$i = 1, 2, \cdots \tau$] and for each integer $n \geq 1$*

$$\left\{ \prod_{i=1}^{\tau} (\delta_i!) \right\} \left| \frac{\prod_{i=1}^{\tau} ((\delta_i n)!)}{(n!)^{\sum_{i=1}^{\tau} \delta_i}} \right.$$

Proof. Apply Theorem 2.10 on each δ_i and multiply. $\qquad\qquad\square$

Theorem 2.12. *For every integer $\delta \geq 2$ and for each integer $n_j \geq 1$ [$j = 1, 2, \cdots \epsilon$]*

$$\{ (\delta!)^{\epsilon} \} \left| \frac{\prod_{j=1}^{\epsilon} (\delta n_j)!}{\prod_{j=1}^{\epsilon} (n_j!)^{\delta}} \right.$$

Proof. Apply Theorem 2.10 on each n_j and multiply. $\qquad\qquad\square$

Theorem 2.13. *For every integer $\delta_i \geq 2$ [$i = 1, 2, \cdots \tau$] and for each integer $n_j \geq 1$ [$j = 1, 2, \cdots \epsilon$]*

$$\left\{ \prod_{i=1}^{\tau} (\delta_i!)^{\epsilon} \right\} \left| \frac{\prod_{j=1}^{\epsilon} \prod_{i=1}^{\tau} \{ (\delta_j n_j)! \}}{\prod_{j=1}^{\epsilon} \left\{ (n_j!)^{\sum_{i=1}^{\tau} \delta_i} \right\}} \right.$$

Proof. Apply Theorem 2.10 on each δ_i and n_j and multiply. $\qquad\qquad\square$

Theorem 2.14. *For every integer $\delta \geq 2$ and for each integer $n \geq 1$, if p is a prime and if [$\cdots$] denotes the greatest integer function, the exponent of the highest power of p which divides*

$$(\delta n)!/(n!)^{\delta} \text{ is } \sum_{h=1}^{\infty} \left\{ [\delta n/p^h] - \delta[n/p^h] \right\}.$$

Proof. Apply the principle of mathematical induction on the variables "n" and "δ". (For $\delta = 2$ we have the standard result.) [1, p.151]. $\qquad\qquad\square$

Theorem 2.15. *For every integer $\delta_i \geq 2$ [$i = 1, 2, \cdots \tau$] and for each integer $n \geq 1$, if p is a prime and if [$\cdots$] denotes the greatest integer function, the exponent of the highest power of p which divides*

$$\frac{\prod_{i=1}^{\tau} ((\delta_i n)!)}{(n!)^{\sum_{i=1}^{\tau} \delta_i}} \text{ is } \sum_{i=1}^{\tau} \sum_{h=1}^{\infty} \left\{ [\delta_i n/p^h] - \delta_i[n/p^h] \right\}.$$

Proof. Apply Theorem 2.14 on each δ_i and add. $\qquad\qquad\square$

Theorem 2.16. *For every integer $\delta \geq 2$ and for each integer $n_j \geq 1$ [$j = 1, 2, \cdots \tau$] if p is a prime and if [$\cdots$] denotes the greatest integer function, the exponent of the highest power of p which divides*

$$\frac{\prod\limits_{j=1}^{\epsilon} (\delta n_j)!}{\prod\limits_{j=1}^{\epsilon} (n_j!)^{\delta}} \ is \ \sum_{j=1}^{\epsilon} \sum_{h=1}^{\infty} \left\{ [\delta n_j / p^h] - \delta[n_j / p^h] \right\}.$$

Proof. Apply Theorem 2.14 on each n_j and add. $\qquad\square$

Theorem 2.17. *For every integer $\delta_i \geq 2$ [$i = 1, 2, \cdots \tau$] and for each integer $n_j \geq 1$ [$j = 1, 2, \cdots \tau$] if p is a prime and if [$\cdots$] denotes the greatest integer function, the exponent of the highest power of p which divides*

$$\frac{\prod\limits_{j=1}^{\epsilon} \prod\limits_{i=1}^{\tau} \left\{ (\delta_j n_j)! \right\}}{\prod\limits_{j=1}^{\epsilon} \left\{ (n_j!)^{\sum\limits_{i=1}^{\tau} \delta_i} \right\}} \ is \ \sum_{j=1}^{\epsilon} \sum_{i=1}^{\tau} \sum_{h=1}^{\infty} \left\{ [\delta_i n_j / p^h] - \delta_i[n_j / p^h] \right\}.$$

Proof. Apply Theorem 2.14 on each δ_i and n_j and add the results. $\qquad\square$

Theorem 2.18. *For every integer $\delta \geq 2$ and for each integer $n \geq 1$, in the prime factorization of $(\delta n)!/(n!)^{\delta}$, the exponent of any prime p such that $(\delta - 1)n < p < \delta n$ is equal to 1. Generalizing we may say that for every integer $\delta \geq 2$ and for each integer $n \geq 1$, in the prime factorization of $(\delta n)!/(n!)^{\delta}$, the exponent of any prime p such that $(\delta - \theta - 1)n < p < (\delta - \theta)n$ is equal to $(\theta + 1)$ [$\theta = 0, 1, \cdots (\delta - 2)$].*

Proof. Apply the principle of mathematical induction on the variables "n" and "δ". (For $\delta = 2$ and $\theta = 0$ have the standard result.) [**1**, p.151]. $\qquad\square$

Theorem 2.19. *For every integer $\delta_i \geq 2$ [$i = 1, 2, \cdots \tau$] and for each integer $n \geq 1$, in the prime factorization of $\dfrac{\prod\limits_{i=1}^{\tau} ((\delta_i n)!)}{(n!)^{\sum\limits_{i=1}^{\tau} \delta_i}}$, the exponent of any prime p such that $\left(\left\{ \prod\limits_{i=1}^{\tau} (\delta_i) \right\} - 1 \right) n < p < \left\{ \prod\limits_{i=1}^{\tau} (\delta_i) \right\} n$ is equal to 1.*

Generalizing we may say that for every integer $\delta_i \geq 2$ [$i = 1, 2, \cdots \tau$] and for each integer $n \geq 1$, in the prime factorization of $\dfrac{\prod\limits_{i=1}^{\tau} ((\delta_i n)!)}{(n!)^{\sum\limits_{i=1}^{\tau} \delta_i}}$, the exponent of any prime p such that $\left(\left\{ \prod\limits_{i=1}^{\tau} (\delta_i) \right\} - \theta - 1 \right) n < p < \left(\left\{ \prod\limits_{i=1}^{\tau} (\delta_i) \right\} - \theta \right) n$ is equal to $(\theta + 1)$ [$\theta = 0, 1, \cdots \left(\left\{ \prod\limits_{i=1}^{\tau} (\delta_i) \right\} - 2 \right)$].

Proof. Apply Theorem 2.18 for every integer $\delta = \left\{ \prod\limits_{i=1}^{\tau} (\delta_i) \right\}$, $(\delta \geq 2)$.   □

Theorem 2.20. *For every integer $\delta \geq 2$ and for each integer $n_j \geq 1$ [j = $1, 2, \cdots \tau$], in the prime factorization of $\dfrac{\prod\limits_{j=1}^{\epsilon} (\delta n_j)!}{\prod\limits_{j=1}^{\epsilon} (n_j!)^{\delta}}$, the exponent of any prime p such   that*
$$(\delta - 1) \left\{ \prod\limits_{j=1}^{\epsilon} (n_j) \right\} < p < \delta \left\{ \prod\limits_{j=1}^{\epsilon} (n_j) \right\} \text{ is equal to } 1.$$
Generalizing we may say that for every integer $\delta \geq 2$ and for each integer $n_j \geq 1$ [j = $1, 2, \cdots \tau$], in the prime factorization of $\dfrac{\prod\limits_{j=1}^{\epsilon} (\delta n_j)!}{\prod\limits_{j=1}^{\epsilon} (n_j!)^{\delta}}$, the exponent of any prime p such that $(\delta - \theta - 1) \left\{ \prod\limits_{j=1}^{\epsilon} (n_j) \right\} < p < (\delta - \theta) \left\{ \prod\limits_{j=1}^{\epsilon} (n_j) \right\}$ is equal to $(\theta + 1)$. [$\theta = 0, 1, \cdots (\delta - 2)$].

Proof. Apply Theorem 2.18 for every integer $n = \left\{ \prod\limits_{j=1}^{\epsilon} (n_j) \right\}$ and $\delta \geq 2$.   □

Theorem 2.21. *For every integer $\delta_i \geq 2$ [i = $1, 2, \cdots \tau$] and for each integer $n_j \geq 1$ [j = $1, 2, \cdots \tau$], in the prime factorization of $\dfrac{\prod\limits_{j=1}^{\epsilon} \prod\limits_{i=1}^{\tau} \{(\delta_j n_j)!\}}{\prod\limits_{j=1}^{\epsilon} \left\{ (n_j!)^{\sum\limits_{i=1}^{\tau} \delta_i} \right\}}$, the exponent of any prime p such that*
$$\left(\left\{ \prod\limits_{i=1}^{\tau} (\delta_i) \right\} - 1 \right) \left\{ \prod\limits_{j=1}^{\epsilon} (n_j) \right\} < p < \left\{ \prod\limits_{i=1}^{\tau} (\delta_i) \right\} \left\{ \prod\limits_{j=1}^{\epsilon} (n_j) \right\}$$
is equal to 1.

Generalizing we may say that for every integer $\delta_i \geq 2$ [i = $1, 2, \cdots \tau$] and for each integer $n_j \geq 1$ [j = $1, 2, \cdots \tau$], in the prime factorization of $\dfrac{\prod\limits_{j=1}^{\epsilon} \prod\limits_{i=1}^{\tau} \{(\delta_j n_j)!\}}{\prod\limits_{j=1}^{\epsilon} \left\{ (n_j!)^{\sum\limits_{i=1}^{\tau} \delta_i} \right\}}$, the exponent of any prime p such that
$$\left(\left\{ \prod\limits_{i=1}^{\tau} (\delta_i) \right\} - \theta - 1 \right) \left\{ \prod\limits_{j=1}^{\epsilon} (n_j) \right\} < p < \left(\left\{ \prod\limits_{i=1}^{\tau} (\delta_i) \right\} - \theta \right) \left\{ \prod\limits_{j=1}^{\epsilon} (n_j) \right\}$$
is equal to $(\theta + 1)$. [$\theta = 0, 1, \cdots \left(\left\{ \prod\limits_{j=1}^{\tau} (\delta_i) \right\} - 2 \right)$].

Proof. Apply Theorem 2.18 for every integer $\delta = \left\{ \prod_{i=1}^{\tau} (\delta_i) \right\}$, $\delta \geq 2$ and for every integer $n = \left\{ \prod_{j=1}^{\epsilon} (n_j) \right\}$. $\qquad \square$

References

[1] David M. Burton; *Elementary Number Theory*, Universal Book Stall, New Delhi, Second Edition. Reprint 1998.

[2] George E. Andrews; *Number Theory*, Hindustan Publishing Corporation (India), Delhi-110007, (1992).

A Remarkable Consequence of the Dirichilet's Theorem

(Dedicated to Smt. Janu, my esteemed teacher at Dr. Nair School, Palghat)

Abstract. A remarkable consequence of Dirichilet's Theorem is stated and proved.

Key Words:- Dirichilet's Theorem, Infinite Primes.
AMS Subject Classification No:- 11A41.

Theorem 0.1. *Let "α" be any positive odd integer whose last digit is not 5. There exists infinite prime numbers that end with α as the last digits.*

Proof. Let α have ξ digits. $[\xi = 1, 2, \cdots \infty]$
Consider the arithmetic progression for each value of ξ

$$\alpha + 10^{\xi}\psi \quad [\psi = 1, 2, \cdots \infty]. \tag{0.1}$$

Each term of the progression closes with the integer α.
α and 10^{ξ} are relatively prime for all values of α and ξ.

Therefore there exist infinite prime numbers that have α as the last digits for each α and each value of ξ by Dirichilet's Theorem applied in each case separately. $\square$

Corollary 0.2. Let x be any positive odd rep-integer whose last digit is not 5. There exists infinite prime numbers that end with x as the last digit. That is there exists infinite prime numbers which end with any number of times any odd integer repeated whose last digit is not 5. These rep-integers are also odd numbers that do not end with 5.

For eg:
Let
$12354354354535435465465465465600000000000005559098098087 = \epsilon$
there exist infinite primes that end with $\epsilon\epsilon\epsilon\epsilon\epsilon\epsilon$ repeated ρ times. $[\rho = 1, 2, 3, \cdots, \infty]$.

The standard results for $\alpha = 1$ and $\alpha = 33$ become special cases of this theorem. The theorem that there exist infinite primes that have a repunit as the last digits of α for each repunit R_{ξ} $[\xi = 1, 2, \cdots]$ is also a special case of this theorem.

References

[1] David M. Burton; *Elementary Number Theory*, Universal Book Stall, New Delhi, Second Edition. Reprint 1998.

Some Fundamental Conjectures on the Distribution of Prime Numbers

(Dedicated to the seamless memory of my beloved parents)

Abstract. Some fundamental conjectures on prime differences are stated.

Key Words:- Prime Numbers, Infinite Generalized Conjectures.
AMS Subject Classification No:- 11A41.

Let us tentatively recall the ancient ageless prime numbers. Let us also remember the great conjectures of Goldbach on prime numbers for their great beauty and fierce vastness.

This brief paper conjectures on the differences of prime numbers.

Conjecture 0.1. Every even number can be expressed as a difference of two prime numbers.

Conjecture 0.2. Every odd number can be expressed as a difference of the sum of two prime numbers and another prime number and also as the difference of a prime number and the sum of two other prime numbers.

Conjecture 0.3. Every even number can be expressed as a difference of two prime numbers in infinite distinct ways.

Let E be any even number and let p and p' be any prime numbers. Then the equation $E = p - p'$ has infinite distinct solutions for every E.

Conjecture 0.4. Every odd number can be expressed as a difference of the sum of two prime numbers and another prime number and also as the difference of a prime number and the sum of two other prime numbers both in infinite distinct ways.

Let O be any odd number and let p, p' and p'' be any prime numbers. Then the equation $O = p + p' - p''$ has infinite distinct solutions for every O.

Let O be any odd number and let p, p' and p'' be any prime numbers. Then the equation $O = p - p' - p''$ has infinite distinct solutions for every O.

Perhaps these conjectures are related to the Goldbach's conjectures. But that is itself another conjecture.

Yet mathematically the conjectures are still unproved for every even and odd number respectively. Some inductive – Proof may be possible.

The conjectures to follow (0.5 and 0.6) surprised me terribly by their infinite vastness and I hope that they are also true.

Conjecture 0.5. Let E be any even number and let $\{p_i\}$ and $\{p_j\}$ be two sets of prime numbers. Let $n_1 - n_2 = \pm 2, 4, 6, \cdots \infty\infty$ for ever ∞. Then the following family of equations has infinite solutions each for all acceptable values of "n_1" and "n_2" in this context.

$$E = \sum_{i=1}^{i=n_1} p_i - \sum_{j=1}^{j=n_2} p_j \quad \boxed{p_i \neq p_j}$$

$n_1 - n_2 = \pm 2, 4, 6, \cdots \infty\infty$ for ever ∞.

Conjecture 0.6. Let O be any odd number and let $\{p_i\}$ and $\{p_j\}$ be two sets of prime numbers. Let $n_1 - n_2 = \pm 1, 3, 5, \cdots \infty\infty$ for ever ∞. Then the following family of equations has infinite solutions each for all acceptable values of "n_1" and "n_2" in this context.

$$O = \sum_{i=1}^{i=n_1} p_i - \sum_{j=1}^{j=n_2} p_j \quad \boxed{p_i \neq p_j}$$

$n_1 - n_2 = \pm 1, 3, 5, \cdots \infty\infty$ for ever ∞.

Insisting on the uniqueness of each p_i and p_j in each case would make the above conjectures more fierce and formidable!

Note:

In these conjectures we have assumed that even the primal number "1" is also a prime number.

Acknowledgements

I first heard of Goldbach's conjectures when I was at IIT Bombay, perhaps from Shaikh Kamran (or was it somebody else?). I must remember Shaji Sebastin for rekindling my interest in Goldbach's conjectures. I also thank Ranganath for reading the paper and pointing out a proof reading error.

References

[1] The ancient Sages and Seers of India: *For Zero and The Decimal System.*
[2] Christian Goldbach;

Epilogue

The amusing thing about most of the papers presented here, is that they are infinite results, in each paper, concealed as formulas! So each of those papers is a tentative "Library of Babel" of Formula-s! I leave it as an exercise to the diligent reader to determine which of these papers are not really infinite in this sense of re-cognition!

Narayanan

www.ingramcontent.com/pod-product-compliance
Lightning Source LLC
Chambersburg PA
CBHW081453170526
45166CB00008B/2416